**NASA
Reference
Publication
1318**

1993

Total Solar Eclipse of 3 November 1994

Fred Espenak
Goddard Space Flight Center
Greenbelt, Maryland

Jay Anderson
Prairie Weather Center
Winnipeg, Manitoba

National Aeronautics and
Space Administration

**Scientific and Technical
Information Branch**

Total Solar Eclipse of 3 Nov 1994

Table of Contents

Eclipse Predictions..1
   Introduction..1
   Path And Visibility...1
   General Maps of the Eclipse Path...2
      Orthographic Projection Map of the Eclipse Path.............................2
      Stereographic Projection Map of the Eclipse Path 3
      Equidistant Conic Projection Maps of the Eclipse Path 3
   Elements, Shadow Contacts and Eclipse Path Tables 3
   Local Circumstances Tables..4
   Detailed Maps of the Umbral Path 5
   Estimating Times of Second And Third Contacts...5
   Mean Lunar Radius...6
   Lunar Limb Profile 7
   Limb Corrections To The Path Limits: Graze Zones....................................8
   Saros History..9
Weather Prospects for the Eclipse 10
   Overview...10
   What Controls the Weather?...10
   Details of the Weather 11
   The Pacific Coast of Peru and Chile...12
   The Bolivian Altiplano and the Andes Mountains..13
   Paraguay and the Gran Chaco...14
   Eastern Paraguay, Northern Argentina, and Brazil 14
   Offshore..15
   Summary...15
Observing the Eclipse 16
   Eye Safety During Solar Eclipses...16
   Sky At Totality 16
   Eclipse Photography...17
   Contact Timings from the Path Limits...18
   Plotting the Path on Maps...19
Algorithms, Ephemerides and Parameters 19
Acknowledgments...19
Bibliography..20
   References...20
   Further Reading...20
Figures...21
Tables 33
Maps of the Umbral Path 61
Request Form for NASA Eclipse Bulletins...69

TOTAL SOLAR ECLIPSE OF 3 NOV 1994

Figures, Tables and Maps

Figures..21
   Figure 1: Orthographic Projection Map of the Eclipse Path....................................23
   Figure 2: Stereographic Projection Map of the Eclipse Path....................................24
   Figure 3: The Eclipse Path in South America..25
   Figure 4: The Eclipse Path in Western South America 26
   Figure 5: The Eclipse Path in Eastern South America 27
   Figure 6: The Lunar Limb Profile At 12:40 UT 28
   Figure 7: Mean Surface Pressure for November..29
   Figure 8: Mean Cloud Cover for November..29
   Figure 9: Frequency of Clear Skies Along the Eclipse Path 30
   Figure 10: The Sky During Totality As Seen At 12:20 UT 31
   Figure 11: The Sky During Totality As Seen At 13:00 UT 32

Tables..33
   Table 1: Elements of the Total Solar Eclipse of 3 Nov 1994...................................35
   Table 2: Shadow Contacts and Circumstances 36
   Table 3: Path of the Umbral Shadow 37
   Table 4: Physical Ephemeris of the Umbral Shadow 38
   Table 5: Local Circumstances on the Center Line 39
   Table 6: Topocentric Data and Path Corrections..40
   Table 7: Mapping Coordinates for the Umbral Path...41
   Table 8: Maximum Eclipse and Circumstances for Argentina, Bolivia and Chile 44
   Table 9: Maximum Eclipse and Circumstances for Brazil......................................46
   Table 10: Maximum Eclipse and Circumstances for Colombia, Equador and
           Venezuela..50
   Table 11: Maximum Eclipse and Circumstances for Paraguay, Peru and Uruguay......52
   Table 12: Maximum Eclipse and Circumstances for Mexico, Central America and
           The Caribbean...54
   Table 13: Maximum Eclipse and Circumstances for Africa 56
   Table 14: Climate Statistics Along the Eclipse Track...58
   Table 15: Solar Eclipse Exposure Guide 59

Maps of the Umbral Path...61
   Map 1: Peru..63
   Map 2: Chile and Bolivia..64
   Map 3: Paraguay 65
   Map 4: Brazil 66
   Map 5: Gough Island..67

Eclipse Predictions

Introduction

On Thursday, 3 November 1994, a total eclipse of the Sun will be visible from the southern half of the Western Hemisphere. The Moon's umbral shadow delineates a path through South America including southern Peru, northern Chile, Bolivia, Paraguay and southern Brazil. The path crosses the South Atlantic and swings south of the African continent with no other landfall except for tiny Gough Island. The path finally ends at sunset in the Indian Ocean south of Madagascar. A partial eclipse will be seen from within the much broader path of the Moon's penumbral shadow including all of South America, southern Mexico and Central America, and most of Africa south of the equator (Figures 1 and 2).

Path And Visibility

The first total eclipse to cross land since July 1991 commences approximately ten hours before the Moon reaches perigee. The path of Moon's umbral shadow begins in the Pacific Ocean about 2000 kilometers west of Peru. As the shadow first contacts Earth along the sunrise terminator (12:02 UT), the path is 135 kilometers wide and the total eclipse lasts 1 minute 52 seconds. Traveling southeast, the umbra quickly makes landfall along the southern coast of Peru at 12:12 UT (Figures 3 and 4). The mysterious sky drawings of Nazca lie just outside the path as the center line parallels the Peruvian coast for the next seven minutes. The duration now lasts 2 and 3/4 minutes and the early morning Sun has an altitude of 27° above the eastern horizon. At an elevation of approximately 2800 meters above sea-level, the city of Arequipa lies near the northern limit and will witness a 58 second total eclipse at 12:15 UT. Down on the coast, Mollendo stands on the center line but frequent marine clouds here may obscure the 2 minutes and 51 seconds of totality. As the path moves inland, it crosses the Peru-Chile border (~12:20 UT) and rapidly gains altitude as it sweeps into the western Andes. Located 1300 meters above the coast, the small town of Putre lies on the center line where the central duration lasts 2 minutes 59 seconds with the Sun at 32°.

Traveling with a surface velocity of 1.402 km/s, the leading edge of the umbra crosses the Chile-Bolivia border even before the trailing edge has entered Chile from Peru. The 175 kilometer wide path through western Bolivia crosses the altiplano, most of which has an elevation of 3000 to 4000 meters and enjoys the driest weather along the entire eclipse path. Unfortunately, the area has few facilities and is difficult to reach. Further east, the silver mining city of Potosí lies 45 kilometers north of the center line but will still enjoy a total phase lasting 2 minutes and 43 seconds. By 12:30 UT, the leading edge of the umbra enters western Paraguay. The duration on the center line is then 3 minutes 19 seconds, the Sun stands 41° above the horizon and the umbra travels with a speed of 1.088 km/s (Figure 5). Continuing southeast, the southern limit skirts Asunción at 12:43 UT where a total eclipse of 41 seconds will be visible. However, observers near the center line can expect a duration of 3 minutes and 40 seconds with a solar altitude of 49°. After briefly crossing a narrow finger of land in northeastern Argentina, the umbra enters southern Brazil at 12:48 UT. Porto Alegre is located 120 kilometers south of the path and will experience a partial eclipse of magnitude 0.966 at 13:00 UT. On the center line, residents of Criciuma will witness a total eclipse of 4 minutes 2 seconds as the Sun stands 59° above the horizon.

After reaching the eastern coastline of Brazil, the shadow heads out across the south Atlantic Ocean where the instant of greatest eclipse[1] occurs at 13:39:06 UT. At that point, the length of totality reaches its maximum duration of 4 minutes 22 seconds, the Sun's altitude is 69°, the path width is 189 kilometers and the umbra's velocity is 0.673 km/s. The remainder of the path crosses open ocean with no further landfall with one minor exception. Gough Island experiences maximum eclipse at 14:29 UT with an umbral duration of 3 minutes 46 seconds while the Sun stands at 53°. Afterwards, the shadow passes 350 kilometers south of South Africa at ~15:10 UT. Cape Town witnesses a tantalizing partial eclipse of magnitude 0.886 at 15:08 UT. Finally, the total eclipse ends at 15:16 UT as the umbra leaves Earth's surface along the sunset terminator in the Indian Ocean about 750 kilometers south of Madagascar. In a period of 3 hours 15 minutes, the Moon's shadow sweeps along a path 14,000 kilometers long, encompassing 0.48 % of Earth's surface area.

---

[1] The instant of greatest eclipse occurs when the distance between the Moon's shadow axis and Earth's geocenter reaches a minimum. Although greatest eclipse differs slightly from the instants of greatest magnitude and greatest duration (for total eclipses), the differences are usually negligible.

GENERAL MAPS OF THE ECLIPSE PATH

ORTHOGRAPHIC PROJECTION MAP OF THE ECLIPSE PATH

Figure 1 is an orthographic projection map of Earth [adapted from Espenak, 1987] showing the path of penumbral (partial) and umbral (total) eclipse. The daylight terminator is plotted for the instant of greatest eclipse with north at the top. The sub-Earth point is centered over the point of greatest eclipse and is marked at **GE** with an asterisk. Earth's sub-solar point at that instant is also indicated by the label **SS**.

The limits of the Moon's penumbral shadow delineate the region of visibility of the partial solar eclipse. This irregular or saddle shaped region often covers more than half of the daylight hemisphere of Earth and consists of several distinct zones or limits. At the northern and/or southern boundaries lie the limits of the penumbra's path. Partial eclipses have only one of these limits, as do central eclipses when the shadow axis falls no closer than about 0.45 radii from Earth's center. Great loops at the western and eastern extremes of the penumbra's path identify the areas where the eclipse begins/ends at sunrise and sunset, respectively. If the penumbra has both a northern and southern limit, the rising and setting curves form two separate, closed loops. Otherwise, the curves are connected in a distorted figure eight. Bisecting the 'eclipse begins/ends at sunrise and sunset' loops is the curve of maximum eclipse at sunrise (western loop) and sunset (eastern loop). The exterior tangency points **P1** and **P4** mark the coordinates where the penumbral shadow first contacts (partial eclipse begins) and last contacts (partial eclipse ends) Earth's surface. If the penumbral path has both a northern and southern limit (as does the November 1994 eclipse), then the interior tangency points **P2** and **P3** are also plotted and correspond to the coordinates where the penumbral cone becomes internally tangent to Earth's disk. Likewise, the points **U1** and **U2** mark the exterior and interior coordinates where the umbral shadow initially contacts Earth (path of total eclipse begins). The points **U3** and **U4** mark the interior and exterior positions of the umbra's final contact with Earth's surface (path of total eclipse ends).

A curve of maximum eclipse is the locus of all points where the eclipse is at maximum at a given time. Curves of maximum eclipse are plotted at each half hour Universal Time (UT). They generally run from the northern to the southern penumbral limits, or from the maximum eclipse at sunrise and sunset curves to one of the limits. The outline of the umbral shadow is plotted every ten minutes in UT. The curves of constant eclipse magnitude[2] delineate the locus of all points where the magnitude at maximum eclipse is constant. These curves run exclusively between the curves of maximum eclipse at sunrise and sunset. Furthermore, they are parallel to the northern/southern penumbral limits and the umbral paths of central eclipses. The northern and southern limits of the penumbra may be thought of as curves of constant magnitude of 0%. The adjacent curves are for magnitudes of 20%, 40%, 60% and 80%. The northern and southern limits of the path of total eclipse are curves of constant magnitude of 100%.

At the top of Figure 1, the Universal Time of geocentric conjunction between the Sun and Moon is given followed by the instant of greatest eclipse. The eclipse magnitude is given for greatest eclipse. For central eclipses (both total and annular), it is equivalent to the geocentric ratio of diameters of the Moon and the Sun. Gamma is the minimum distance of the Moon's shadow axis from Earth's center in units of equatorial Earth radii. The shadow axis passes south of Earth's geocenter for negative values of Gamma. Finally, the extrapolated value of ΔT[3] used in the calculations is given.

---

[2] Eclipse magnitude is defined as the fraction or percentage of the Sun's diameter occulted by the Moon. It's usually expressed at greatest eclipse. Eclipse magnitude is strictly a ratio of *diameters* and should not be confused with eclipse obscuration which is a measure of the Sun's surface *area* occulted by the Moon.

[3] ΔT is the difference between Terrestrial Dynamical Time and Universal Time

STEREOGRAPHIC PROJECTION MAP OF THE ECLIPSE PATH

The stereographic projection of Earth in Figure 2 depicts the path of penumbral and umbral eclipse in greater detail. The map is oriented with the point of greatest eclipse near the center and north is at the top. International political borders are shown and circles of latitude and longitude at plotted at 20° increments. The saddle shaped region of penumbral or partial eclipse includes labels identifying the northern and southern limits, curves of eclipse begins or ends at sunrise, curves of eclipse begins or ends at sunset, and curves of maximum eclipse at sunrise and sunset. Curves of constant eclipse magnitude are plotted for 20%, 40%, 60% and 80%, as are the limits of the path of total eclipse. Also included are curves of greatest eclipse for every thirty minutes Universal Time.

Figures 1 and 2 may be used to quickly determine the approximate time and magnitude of greatest eclipse for any location from which the eclipse is visible.

EQUIDISTANT CONIC PROJECTION MAPS OF THE ECLIPSE PATH

Figures 3, 4, and 5 are equidistant conic projection maps which isolate specific regions of the eclipse path. The projection was selected to minimize distortion over the regions depicted. Once again, curves of maximum eclipse and constant eclipse magnitude are plotted along with identifying labels. A linear scale is included for estimating approximate distances (kilometers) in each figure. Within the northern and southern limits of the path of totality, the outline of the umbral shadow is plotted at ten minute intervals. Figure 3 is drawn at a scale of ~1:31,800,000, while figures 4 and 5 are drawn at a scale of ~1:8,560,000. All three figures include the positions of many of the larger cities or metropolitan areas in and near the central path. The size of each city is logarithmically proportional to its population according to 1990 census data. The last two figures also include the center line of the umbral path.

ELEMENTS, SHADOW CONTACTS AND ECLIPSE PATH TABLES

The geocentric ephemeris for the Moon and Sun, various parameters and constants used in the predictions, the besselian elements (polynomial form) are given in Table 1. The eclipse predictions and elements were derived from solar and lunar data contained in the DE200 and LE200 ephemerides developed jointly by the Jet Propulsion Laboratory and the U. S. Naval Observatory for use in the *Astronomical Almanac* for 1984 and after. Unless otherwise stated, all predictions are based on center of mass positions for the Sun and Moon with no corrections made for center of figure, center of motion, lunar limb profile or atmospheric refraction. Furthermore, these predictions depart from IAU convention by using a smaller constant for the mean lunar radius k for all umbral contacts (see: LUNAR LIMB PROFILE). Times are expressed in either Terrestrial Dynamical Time (TDT) or in Universal Time (UT), where the best value of ΔT available at the time of preparation is used.

Table 2 lists all external and internal contacts of penumbral and umbral shadows with Earth. They include TDT times and geodetic coordinates both with and without corrections for ΔT. These contacts are defined as follows:

P1 - Instant of first external tangency of penumbral shadow cone with Earth's limb.
(partial eclipse begins)
P2 - Instant of first internal tangency of penumbral shadow cone with Earth's limb.
P2 - Instant of last internal tangency of penumbral shadow cone with Earth's limb.
P4 - Instant of last external tangency of penumbral shadow cone with Earth's limb.
(partial eclipse ends)

U1 - Instant of first external tangency of umbral shadow cone with Earth's limb.
(umbral eclipse begins)
U2 - Instant of first internal tangency of umbral shadow cone with Earth's limb.
U2 - Instant of last internal tangency of umbral shadow cone with Earth's limb.
U4 - Instant of last external tangency of umbral shadow cone with Earth's limb.
(umbral eclipse ends)

Similarly, the northern and southern extremes of the penumbral and umbral paths, and extreme limits of the umbral center line are given. The IAU longitude convention is used throughout this publication (i.e. - eastern longitudes are positive; western longitudes are negative; negative latitudes are south of the Equator).

The path of the umbral shadow is delineated at five minute intervals in Universal Time in Table 3. Coordinates of the northern limit, the southern limit and the center line are listed to the nearest tenth of an arc-minute (~185 m at the Equator). The path azimuth, path width and umbral duration are calculated for the center line position. The path azimuth is the direction of the umbral shadow's motion projected onto the surface of the Earth. Table 4 presents a physical ephemeris for the umbral shadow at five minute intervals in UT. The center line coordinates are followed by the topocentric ratio of the apparent diameters of the Moon and Sun, the eclipse obscuration[4], and the Sun's altitude and azimuth at that instant. The central path width, the umbral shadow's major and minor axes and its instantaneous velocity with respect to Earth's surface are included. Finally, the center line duration of the umbral phase is given.

Local circumstances for each center line position listed in Tables 3 and 4 are presented in Table 5. The first three columns give the Universal Time of maximum eclipse, the center line duration of totality and the altitude of the Sun at that instant. The following columns list each of the four eclipse contact times followed by their related contact position angles and the corresponding altitude of the Sun. The four contacts[5] identify significant stages in the progress of the eclipse. The position angles P and V identify the point along the Sun's disk where each contact occurs[6]. The altitude of the Sun at second and third contact is omitted since it is always within 1° of the altitude at maximum eclipse (column 3).

Table 6 presents topocentric values at maximum eclipse for the Moon's horizontal parallax, semi-diameter, relative angular velocity with respect to the Sun, and libration in longitude. The altitude and azimuth of the Sun are given along with the azimuth of the umbral path. The northern limit position angle identifies the point on the lunar disk defining the umbral path's northern limit. It is measured counter-clockwise from the north point of the lunar disk. In addition, corrections to the path limits due to the lunar limb profile are listed. The irregular profile of the Moon results in a zone of 'grazing eclipse' at each limit which is delineated by interior and exterior contacts of lunar features with the Sun's limb. The section LIMB CORRECTIONS TO THE PATH LIMITS: GRAZE ZONES describes this geometry in greater detail. Corrections to the center line durations due to the lunar limb profile are also included. When added to the durations in Tables 3, 4, 5 and 7, a slightly shorter central total phase is predicted.

To aid and assist in the plotting of the umbral path on large scale maps, the path coordinates are also tabulated at 1° intervals in longitude in Table 7. The latitude of the northern limit, southern limit and center line for each longitude is tabulated along with the Universal Time of maximum eclipse at each position. Finally, local circumstances on the center line at maximum eclipse are listed and include the Sun's altitude and azimuth, the umbral path width and the central duration of totality.

LOCAL CIRCUMSTANCES TABLES

Local circumstances from over 300 cities, metropolitan areas and places in South America, Central America and Africa are presented in Tables 8 through 13. Each table is broken down into two parts. The first part, labeled a, appears on even numbered pages and gives circumstances at maximum eclipse[7] for each location. The coordinates are listed along with the location's elevation (meters) above sea-level, if known. If the elevation is unknown (i.e. - not in the data base), then the local circumstances for that location are calculated at sea-level. In any case, the elevation does not play a significant role in the predictions unless the location is near the umbral path limits and the Sun's altitude is relatively small (<15°). The Universal Time of maximum eclipse (either partial or total) is listed to an accuracy of 0.1 seconds. If the eclipse is total, then the umbral duration and the path width are given. Next, the altitude and azimuth of the Sun at maximum eclipse are listed along with the position angles P and V of the Moon's disk with respect to the Sun. Finally, the magnitude and obscuration are listed at the instant of maximum eclipse. Note that for umbral eclipses (annular and total), the eclipse magnitude is identical to the topocentric ratio of the Moon's

---

[4] Eclipse obscuration is defined as the fraction of the Sun's surface area occulted by the Moon.

[5] First contact is defined as the instant of external tangency between the Sun and Moon; it marks the beginning of the partial eclipse.
Second and third contacts define the two instants of internal tangency between the Sun and Moon; they signify the commencement and termination of the umbral (total or annular) phase.
Fourth contact is the instant of last external contact and it marks the end of the partial eclipse.

[6] P is defined as the contact angle measured counter-clockwise from the *north* point of the Sun's disk.
V is defined as the contact angle measured counter-clockwise from the *zenith* point of the Sun's disk.

[7] For partial eclipses, maximum eclipse is the instant when the greatest fraction of the Sun's diameter is occulted. For umbral eclipses (total or annular), maximum eclipse is the instant of mid-totality or mid-annularity.

and Sun's apparent diameters. Furthermore, the eclipse magnitude is always less than 1 for annular eclipses and equal to or greater than 1 for total eclipses.

The second part of each table, labeled **b**, is found on odd numbered pages. It gives local circumstances for each location listed on the facing page at each contact during the eclipse. The Universal Time of each contact is given along with the altitude of the Sun, followed by position angles **P** and **V**. These angles identify the point along the Sun's disk where each contact occurs and are measured counter-clockwise from the north and zenith points, respectively. Locations outside the umbral path miss the umbral eclipse and only witness first and fourth contacts. The effects of refraction have been included in these calculations although no correction has been applied for center of figure or the lunar limb profile.

Locations were chosen based on position near the central path, general geographic distribution and population. The primary source for geographic coordinates is *The New International Atlas* (Rand McNally, 1991). Elevations for major cities were taken from *Climates of the World* (U. S. Dept. of Commerce, 1972). In this rapidly changing political world, it is often difficult to ascertain the correct name or spelling for a given location. Therefore, the information presented here is for location purposes only and is not meant to be authoritative. Furthermore, it does not imply recognition of status of any location by the United States Government. Corrections to names, spellings, coordinates and elevations is solicited in order to update the geographic data base for future eclipse predictions.

DETAILED MAPS OF THE UMBRAL PATH

The path of totality has been plotted by hand on a set of five detailed maps appearing in the last section of this publication. The maps are Global Navigation and Planning Charts or GNC's from the Defense Mapping Agency which use a Lambert conformal conic projection. More specifically, GNC-18 covers the South American section of the path while GNC-17 covers the South Atlantic (Gough Island). GNC's have a scale of 1:5,000,000 (1 inch ~ 69 nautical miles), which is adequate for showing major cities, highways, airports, rivers, bodies of water and basic topography required for eclipse expedition planning including site selection, transportation logistics and weather contingency strategies.

Northern and southern limits as well as the center line of the path are drawn using predictions from Table 3. No corrections have been made for center of figure or lunar limb profile. However, such corrections have little or no effect at this scale. Although, atmospheric refraction has not been included, its effects play a significant role only at low solar altitudes (<15°). In any case, refraction corrections to the path are uncertain since they depend on the atmospheric temperature-pressure profile which cannot be predicted in advance. If observations from the graze zones are planned, then the path must be plotted on higher scale maps using limb corrections in Table 6. See PLOTTING THE PATH ON MAPS for sources and more information. The GNC paths also depict the curve of maximum eclipse at five minute increments in Universal Time [Table 3].

ESTIMATING TIMES OF SECOND AND THIRD CONTACTS

The times of second and third contact for any location not listed in this publication can be estimated using the detailed maps found in the final section. Alternatively, the contact times can be estimated from maps on which the umbral path has been plotted. Table 7 lists the path coordinates conveniently arranged in 1° increments of longitude to assist plotting by hand. The path coordinates in Table 3 define a line of maximum eclipse at five minute increments in time. These lines of maximum eclipse each represent the projection diameter of the umbral shadow at the given time. Thus, any point on one of these lines will witness maximum eclipse (i.e.: mid-totality) at the same instant. The coordinates in Table 3 should be added to the map in order to construct lines of maximum eclipse.

The estimation of contact times for any one point begins with an interpolation for the time of maximum eclipse at that location. The time of maximum eclipse is proportional to a point's distance between two adjacent lines of maximum eclipse, measured along a line parallel to the center line. This relationship is valid along most of the path with the exception of the extreme ends, where the shadow experiences its largest acceleration. The center line duration of totality **D** and the path width **W** are similarly interpolated from the values of the adjacent lines of maximum eclipse as listed in Table 3. Since the location of interest probably does not lie on the center line, it is useful to have an expression for calculating the duration of totality **d** as a function of its perpendicular distance **a** from the center line:

$$d = D \left(1 - (2a/W)^2\right)^{1/2} \text{ seconds} \qquad [1]$$

where: D = duration of totality on the center line (seconds)
W = width of the path (kilometers)
a = perpendicular distance from the center line (kilometers)

If t_m is the interpolated time of maximum eclipse for the location, then the approximate times of second and third contacts (t_2 and t_3, respectively) are:

Second Contact: $\qquad t_2 = t_m - d/2 \qquad [2]$
Third Contact: $\qquad t_3 = t_m + d/2 \qquad [3]$

The position angles of second and third contact (either P or V) for any location off the center line are also useful in some applications. First, linearly interpolate the center line position angles of second and third contacts from the values of the adjacent lines of maximum eclipse as listed in Table 5. If X_2 and X_3 are the interpolated center line position angles of second and third contacts, then the position angles x_2 and x_3 of those contacts for an observer located a kilometers from the center line are:

Second Contact: $\qquad x_2 = X_2 - \text{ArcSin}(2a/W) \qquad [4]$
Third Contact: $\qquad x_3 = X_3 + \text{ArcSin}(2a/W) \qquad [5]$

where: X_n = the interpolated position angle (either P or V) of contact n on center line
x_n = the interpolated position angle (either P or V) of contact n at location
D = duration of totality on the center line (seconds)
W = width of the path (kilometers)
a = perpendicular distance from the center line (kilometers)
(use negative values for locations south of the center line)

MEAN LUNAR RADIUS

A fundamental parameter used in the prediction of solar eclipses is the Moon's mean radius k, expressed in units of Earth's equatorial radius. The actual radius of the Moon varies as a function of position angle and libration due to the irregularity of the lunar limb profile. From 1968 through 1980, the Nautical Almanac Office used two separate values for k in their eclipse predictions. The larger value (k=0.2724880), representing a mean over lunar topographic features, was used for all penumbral (i.e. - exterior) contacts and for total eclipses. A smaller value (k=0.272281), representing a mean minimum radius, was reserved exclusively for umbral (i.e. - interior) contact calculations of total eclipses [*Explanatory Supplement*, 1974]. Unfortunately, the use of two different values of k for umbral eclipses introduces a discontinuity in the case of hybrid or annular-total eclipses.

In August 1982, the IAU General Assembly adopted a value of k=0.2725076 for the mean lunar radius. This value is currently used by the Nautical Almanac Office for all solar eclipse predictions [Fiala and Lukac, 1983] and is believed to be the best mean radius, averaging mountain peaks and low valleys along the Moon's rugged limb. In general, the adoption of one single value for k is commendable because it eliminates the discontinuity in the case of annular-total eclipses and ends confusion arising from the use of two different values. However, the use of even the best 'mean' value for the Moon's radius introduces a problem in predicting the character and duration of umbral eclipses, particularly total eclipses. A total eclipse can be defined as an eclipse in which the Sun's disk is completely occulted by the Moon. This cannot occur so long as any photospheric rays are visible through deep valleys along the Moon's limb [Meeus, Grosjean and Vanderleen, 1966]. But the use of the IAU's mean k guarantees that some annular or annular-total eclipses will be misidentified as total. A case in point is the eclipse of 3 October 1986. The *Astronomical Almanac* identified this event as a total eclipse of 3 seconds duration when in it was in fact a beaded annular eclipse. Clearly, a smaller value of k is needed since it is more representative of the deepest lunar valley floors, hence the minimum solid disk radius and ensures that an eclipse is truly total.

Of primary interest to most observers are the times when central eclipse begins and ends (second and third contacts, respectively) and the duration of the central phase. When the IAU's mean value for k is used to calculate these times, they must be corrected to accommodate low valleys (total) or high mountains (annular) along the Moon's limb. The calculation of these corrections is not trivial but must be performed, especially if one plans to observe near the path limits [Herald, 1983]. For observers near the center line of a total eclipse, the limb corrections can be closely approximated by using a smaller value of **k** which accounts for the valleys along the profile.

This work uses the IAU's accepted value of **k** (k=0.2725076) for all penumbral (exterior) contacts. In order to avoid eclipse type misidentification and to predict central durations which are closer to the actual durations observed at total eclipses, we depart from convention by adopting the smaller value for **k** (k=0.272281) for all umbral (interior) contacts. This is consistent with predictions published in *Fifty Year Canon of Solar Eclipses: 1986 - 2035* [Espenak, 1987]. Consequently, the smaller **k** produces shorter umbral durations and narrower paths for total eclipses when compared with calculations using the IAU value for **k**. Similarly, the smaller **k** predicts longer umbral durations and wider paths for annular eclipses.

LUNAR LIMB PROFILE

Eclipse contact times, the magnitude and the duration of totality all ultimately depend on the angular diameters and relative velocities of the Sun and the Moon. Unfortunately, these calculations are limited in accuracy by the departure of the Moon's limb from a perfectly circular figure. The Moon's surface exhibits a rather dramatic topography which manifests itself as an irregular limb when seen in profile. Most eclipse calculations assume some mean lunar radius which averages high mountain peaks and low valleys along the Moon's rugged limb. Such an approximation is acceptable for many applications, but if higher accuracy is needed, the Moon's actual limb profile must be considered. Fortunately, an extensive body of knowledge exists on this subject in the form of Watts' limb charts [Watts, 1963]. These data are the product of a photographic survey of the marginal zone of the Moon and give limb profile heights with respect to an adopted smooth reference surface (or datum). Analyses of lunar occultations of stars by Van Flandern [1970] and Morrison [1979] have shown that the average cross-section of Watts' datum is slightly elliptical rather than circular. Furthermore, the implicit center of the datum (i.e. - the center of figure) is displaced from the Moon's center of mass. In a follow-up analysis of 66000 occultations, Morrison and Appleby [1981] have found that the radius of the datum appears to vary with libration. These variations produce systematic errors in Watts' original limb profile heights which attain 0.4 arc-seconds at some position angles. Thus, corrections to Watts' limb profile data are necessary to ensure that the reference datum is a sphere with its center at the center of mass.

The Watts charts have been digitized by Her Majesty's Nautical Almanac Office in Herstmonceux, England, and transformed to grid-profile format at the U. S. Naval Observatory. In this computer readable form, the Watts limb charts lend themselves to the generation of limb profiles for any lunar libration. Ellipticity and libration corrections may be applied to refer the profile to the Moon's center of mass. Such a profile can then be used to correct eclipse predictions which have been generated using a mean lunar limb.

Along the eclipse path, the Moon's topocentric libration (physical + optical libration) in longitude ranges from l= 0.0° to l=−1.6°. Thus, a limb profile with the appropriate libration is required in any detailed analysis of contact times, central durations, etc.. Nevertheless, a profile with an intermediate libration is valuable for general planning for any point along the path. The center of mass corrected lunar limb profile presented in Figure 6 is for the center line at 12:40 UT. At that time, the Moon's topocentric librations are l=−0.30°, b=+0.10° and c=+19.65°, and the apparent topocentric semi-diameters of the Sun and Moon are 967.4 and 1015.6 arc-seconds respectively. The Moon's angular velocity with respect to the Sun is 0.446 arc-seconds per second.

The radial scale of the profile in Figure 6 (see scale to upper left) is greatly exaggerated so that the true limb's departure from the mean lunar limb is readily apparent. The mean limb with respect to the center of figure of Watts' original data is shown along with the mean limb with respect to the center of mass. Note that all the predictions presented in this paper are calculated with respect to the latter limb unless otherwise noted. Position angles of various lunar features can be read using the protractor in the center of the diagram. The position angles of second and third contact are clearly marked along with the north pole of the Moon's axis of rotation and the observer's zenith at mid-totality. The dashed line arrows identify the points on the limb which define the northern and southern limits of the path. To the upper left of the profile are the Moon's mean lunar radius **k** (expressed in Earth equatorial radii), topocentric semi-diameter **SD** and horizontal parallax **HP**. As discussed in the section MEAN LUNAR RADIUS, the Moon's mean radius **k** (k=0.2722810) is smaller than the adopted IAU value (k=0.2725076). To the upper right of the profile are the Sun's semi-diameter **SUN SD**, the angular velocity of the Moon with respect to the Sun **VELOC.** and the position angle of the path's northern/southern limit axis **LIMITS**. In the lower right are the Universal Times of the four contacts and maximum eclipse. The geographic coordinates and local circumstances at maximum eclipse are given along the bottom of the figure.

In investigations where accurate contact times are needed, the lunar limb profile can be used to correct the nominal or mean limb predictions. For any given position angle, there will be a high mountain (annular eclipses) or a low valley (total eclipses) in the vicinity which ultimately determines the true instant of contact. The difference, in time, between the Sun's position when tangent to the contact point on the mean limb and tangent to the highest mountain (annular) or lowest valley (total) at actual contact is the desired correction to the predicted contact time. On the exaggerated radial scale of Figure 6, the Sun's limb can be represented as an epicyclic curve which is tangential to the mean lunar limb at the point of contact and departs from the limb by **h** as follows:

$$h = S (m-1) (1-\cos[C]) \qquad [6]$$

where: S = the Sun's semi-diameter
m = the eclipse magnitude
C = the angle from the point of contact

Herald [1983] has taken advantage of this geometry to develop a graphical procedure for estimating correction times over a range of position angles. Briefly, a displacement curve of the Sun's limb is constructed on a transparent overlay by way of equation [6]. For a given position angle, the solar limb overlay is moved radially from the mean lunar limb contact point until it is tangent to the lowest lunar profile feature in the vicinity. The solar limb's distance **d** (arc-seconds) from the mean lunar limb is then converted to a time correction Δ by:

$$\Delta = d \, v \, \cos[X - C] \qquad [7]$$

where: d = the distance of Solar limb from mean lunar limb (arc-sec)
v = the angular velocity of the Moon with respect to the Sun (arc-sec/sec)
X = the center line position angle of the contact
C = the angle from the point of contact

This operation may be used for predicting the formation and location of Baily's beads. When calculations are performed over large range of position angles, a contact time correction curve can then be constructed.

Since the limb profile data are available in digital form, an analytic solution to the problem is possible which is straight forward and quite robust. Curves of corrections to the times of second and third contact for most position angles have been computer generated and are plotted in Figure 6. In interpreting these curves, the circumference of the central protractor functions as the nominal or mean contact time (using the Moon's mean limb) as a function of position angle. The departure of the correction curve from the mean contact time can then be read directly from Figure 6 for any position angle by using the radial scale in the upper right corner (units in seconds of time). Time corrections external to the protractor (most second contact corrections) are added to the mean contact time; time corrections internal to the protractor (all third contact corrections) are subtracted from the mean contact time.

Across all of South America, the Moon's topocentric libration in longitude at maximum eclipse is within 0.2° of its value at 12:40 UT. Therefore, the limb profile and contact correction time curves in Figure 6 may be used in all but the most critical investigations.

LIMB CORRECTIONS TO THE PATH LIMITS: GRAZE ZONES

The northern and southern umbral limits provided in this publication were derived using the Moon's center of mass and a mean lunar radius. They have not been corrected for the Moon's center of figure or the effects of the lunar limb profile. In applications where precise limits are required, Watts' limb data must be used to correct the nominal or mean path. Unfortunately, a single correction at each limit is not possible since the Moon's libration in longitude and the contact points of the limits along the Moon's limb each vary as a function of time and position along the umbral path. This makes it necessary to calculate a unique correction to the limits at each point along the path. Furthermore, the northern and southern limits of the umbral path are actually paralleled by a relatively narrow zone where the eclipse is neither penumbral nor umbral. An observer positioned here will witness a solar crescent which is fragmented into a series of bright beads and short segments whose morphology changes quickly with the rapidly varying geometry of the Moon with respect to the Sun. These beading phenomena are caused by the appearance of photospheric rays which alternately pass through deep lunar valleys and hide behind high mountain peaks as the Moon's irregular limb grazes the edge of the Sun's disk. The geometry is directly analogous to the case of grazing occultations of stars by the Moon. The graze zone is typically five to ten kilometers wide and its interior and exterior boundaries can be predicted using the lunar limb profile. The interior boundaries define the actual limits of the umbral eclipse (both total and annular) while the exterior boundaries set the outer limits of the grazing eclipse zone.

Table 6 provides topocentric data and corrections to the path limits due to the true lunar limb profile. At five minute intervals, the table lists the Moon's topocentric horizontal parallax, the semi-diameter, the relative angular velocity of the Moon with respect to the Sun and lunar libration in longitude. The center line altitude and azimuth of the Sun is given, followed by the azimuth of the umbral path. The position angle of the point on the Moon's limb which defines the northern limit of the path is measured counter-clockwise (i.e. - eastward) from the north point on the limb. The path corrections to the northern and southern limits are listed as interior and exterior components in order to define the graze zone. Positive corrections are in the northern sense while negative shifts are in the southern sense. These corrections [minutes of arc in latitude] may be added directly to the path coordinates listed in Table 3. Corrections to the center line umbral durations due to the lunar limb profile are also included and they are all negative. Thus, when added to the central durations given in Tables 3, 4, 5 and 7, a slightly shorter central total phase is predicted.

SAROS HISTORY

The total eclipse of 3 November 1994 is the forty-fourth member of Saros series 133, as defined by van den Bergh (1955). All eclipses in the series occur at the Moon's ascending node and gamma[8] decreases with each member in the series. The family began on 13 July 1219 with a partial eclipse in the northern hemisphere. During the next two centuries, a dozen partial eclipses occurred with the eclipse magnitude of each succeeding event gradually increasing. Finally, the first umbral eclipse occurred on 20 November 1435. The event was an annular eclipse with no northern limit. The followed five eclipses were also annular with maximum umbral durations decreasing from 74 to 7 seconds. The nineteenth event occurred on 24 January 1544 and was of annular/total nature. From the mid-sixteenth through mid-nineteenth centuries, the series continued to produce total eclipses with monotonically increasing durations. This trend culminated with the total eclipse of 7 August 1850 which passed through the Hawaiian Islands and had a maximum duration of 6 minutes 50 seconds.

While Saros 133 has continued to produce total eclipses throughout the twentieth century, the duration of each succeeding event is now decreasing as Earth moves progressively closer to perihelion. The most recent eclipse of the series took place on 23 October 1976. It was visible along the southeastern coast of Australia but the maximum umbral duration of 4 minutes 46 seconds occurred over the Indian Ocean. In comparison, the maximum duration of the 3 November 1994 event is 4 minutes 23 seconds in the South Atlantic. The next eclipse of the series will be 13 November 2012. While the umbra crosses northern Australia, most of the path lies over the South Pacific where the maximum of 4 minutes 2 seconds takes place.

During the next 150 years, each path moves deeper into the southern hemisphere as the maximum duration gradually decreases, dropping below the three minute mark with the eclipse of 8 January 2103. However, the eclipses from 21 February 2175 through 28 April 2283 exhibit a monotonic increase in duration from 2 minutes 50 seconds to 3 minutes 13 seconds. This century long reversal of the decreasing trend is due primarily to the passage of Earth through the vernal equinox. The effect briefly shifts the southerly migrating eclipse paths back towards the equator where the larger rotational velocity extends the duration of totality. The duration drops once more with the last five central eclipses of the series. The final total eclipse occurs on 21 June 2373 with a duration of 1 minute 24 seconds. As the series winds down, the first of seven remaining partial eclipses occurs on 3 July 2391 and exhibits a magnitude of 0.867 from high southerly latitudes. Saros 133 finally ends with its seventy-second event, the partial eclipse of 5 September 2499.

In summary, Saros series 133 includes 72 eclipses with the following distribution:

Saros 133	*Partial*	*Annular*	*Ann/Total*	*Total*
Non-Central	19	0	0	0
Central	—	6	1	46

---

[8] Gamma is measured in Earth radii and is the minimum distance of the Moon's shadow axis from Earth's center during an eclipse. This occurs at and defines the instant of greatest eclipse. Gamma takes on negative values when the shadow axis is south of the Earth's center.

Weather Prospects for the Eclipse

Overview

The jumble of terrain in South America divides the eclipse track into many varieties of weather. Peruvian coastal plains that cling to the continental edge soar to the high mountain plateaus of Bolivia within a distance of only two hundred kilometers. From the height of the plateau the shadow path descends through winding mountain valleys into swampy foothills to cross the rolling hills and river valleys of Paraguay. Across southern Brazil, a vestigial mountain range challenges the Moon's shadow before the track splashes into the waters of the Atlantic Ocean.

The eclipse observer can select his or her spot according to individual circumstances and willingness for adventuresome travel. Cool windy beaches, freezing plains, tropical forests, prairie grasslands or subtropical cities - this eclipse covers them all! And for those infected by a desire to travel to lost destinations, a tiny Atlantic island off the coast of South Africa offers a final land's end's view of the spectacle before the shadow heads back into space.

What Controls the Weather?

Weather is a product of winds and moisture. During this eclipse, the winds are controlled by two large oceanic high pressure systems or anticyclones (Figure 7). One of these lies west of Chile. The other is found off the coast of South Africa in the South Atlantic. These highs are permanent residents of the mid-ocean. Though they wax and wane with each passing disturbance, they never disappear completely.

A weaker continental low lying east of the Andes over Brazil divides the two highs. This low is created by warmer temperatures over the land, and is easily displaced by stronger but more transient weather systems that approach from the Atlantic. Because of the low, the weather across the eastern part of South America is more variable than that west of the Andes. The Pacific high and the Brazilian low regulate the weather over the land portion of this eclipse. The Atlantic high, lying closer to the African coast, has little influence over South America, but is the dominating system over the Atlantic portion of track.

In the southern hemisphere, winds circulate in a counterclockwise direction around the high pressure cells. On the Pacific side this brings a stubborn southerly wind onto the Peruvian coast. This wind is further strengthened by local sea breezes that build the largest sand dunes in the world. Sea breezes are onshore winds caused by the heating of the land along the coast. The heated air rises upward and cooler ocean air then moves inland to replace it.

The influence of the highs extends beyond directing wind flow. Air inside anticyclones descends from higher levels in the atmosphere. Air which is descending is warmed and dried. At the ocean surface is the cold Peruvian Current, bringing water northward from the sub polar regions off Antarctica. Warm air aloft and cool temperatures below combine to create a temperature inversion. The marine inversion off the Peruvian coast is very similar to its cousin off the coast of California. Both are very persistent.

Inversions are stable phenomena because cold air is heavy and sinks to the bottom of the atmosphere, resisting mixing with warmer layers above. They are of serious concern because marine inversions also collect water vapor from the ocean surface and become very cloudy. The water vapor is trapped, unable to mix with drier parts of the atmosphere above. The cloud isn't very deep, usually no more than about 900 meters (though occasionally up to 1500), but it is very resistant to clearing. In the air above the inversion, brilliant sunshine is the rule. Because the marine cloud is so shallow, it contains too little moisture to bring rain. What moisture is available comes from a persistent drizzle or a heavy morning dew that feeds specially adapted plants. Not much precipitation accumulates, resulting in a climatology that is both cloudy and dry. When it does rain along the Peruvian coast, it usually falls on the slopes of the Andes above the inversion.

About every three to seven years a phenomenon known as El Niño develops along the Peruvian coast. This is a complex oscillation in the weather patterns across the equatorial Pacific which reverses the normal climatology of the tropics when ocean currents warm and trade winds weaken. During El Niño years, the coasts of Ecuador and northern Peru typically become very wet and cloudy while droughts develop over southern Peru. An El Niño began in late 1991 and weakened in the summer of 1992. It was an unusual El Niño and did not completely fit the weather patterns which usually come with this oscillation. Its demise was also ambiguous and there were signs that the pattern was lingering into the spring of 1993. Since the El Niño occurred in 1991-92 and possibly into 1993, it would be unusual for another to develop in time for the November 1994 eclipse. Most likely normal weather and cloud patterns will prevail.

East of the Andes, the low pressure system that resides over Brazil draws air inland from the Atlantic Ocean. Air flow from the Pacific is blocked by the mountains. Storms and disturbances approach from the east. November is late spring in the southern hemisphere, comparable to May in the north. Summer weather patterns are not fully established, but the winter is gone, taking most of the storms and blustery weather. The eclipse takes place during a time of transition, but the weather may be quite distinctive for the season, and not just an average of the summer and winter patterns. In particular, November is the sunniest month over some parts of the eclipse track.

Above the surface, upper level winds blow mostly from the west except in Peru where they are more variable and occasionally from eastward. Because they are blowing out of the Pacific anticyclone, these upper westerlies are usually dry and cloud-free. Occasional weak disturbances will bring scattered showers and thundershowers to the Pacific side of the Andes, but these are much more likely on the other side of the mountain range, in eastern Bolivia.

Figure 8 shows the large scale pattern of the cloud cover along the eclipse track. This chart was constructed from 10 years of satellite data analyzed by Russian researchers[9] at St. Petersburg (once Leningrad) University. To calculate the mean amount of cloud, the Earth was divided into large blocks of latitude and longitude. By averaging over a large area, the details of the cloudiness have been lost. In particular, Figure 8 gives a poor representation of cloud patterns over the mountains but those over eastern South America and the oceans are well shown. The most notable features are the high levels of cloudiness over the oceans and the sunnier conditions over Paraguay.

Figure 9 is a much more detailed look at cloud cover along the eclipse track. To make this graph, the eclipse track was plotted on daily satellite pictures taken near eclipse time during late October and most of November in 1991 and 1992. Cloud cover was then estimated at each degree of longitude along the track and assembled into the graph.

During the 1991-92 El Niño, abnormally wet weather covered parts of Argentina, Paraguay, Bolivia and Brazil from mid-November 1991 to January 1992. Since Figure 9 includes approximately one week of data from this period, the graph will show some biases which distort the relationship between the regions. In particular, Peru and the altiplano may be slightly sunnier than would otherwise be the case while more easterly sections probably show a little too much cloud. In November 1992, weather patterns seemed to have returned to those typical of non-El Niño years. Figure 9 should be used cautiously though the general relationship between regions is probably reliable. Figure 8 (based on a decade of satellite data) is more faithful to the large scale climatological cloud cover. Longer term statistics (which also include El Niño years) can be found in Table 14.

DETAILS OF THE WEATHER

The track can be divided into four weather regions, ordained mostly by the Andes mountains, and partially by the coastal mountains in Brazil. From west to east these are:

1) the Pacific coast of Peru and Chile
2) the mountains and altiplano of Bolivia
3) the Gran Chaco, Paraguay and Argentina
4) the hills and mountains of southeast Brazil.

In the southern hemisphere, winds blow clockwise around lows and counter-clockwise around highs. The massive ramparts of the Andes deflect the winds from their usual direction and turn them into the many valleys and ridges of the mountains. Each valley comes with its own winds, sometimes enhancing or retarding the larger pattern. This wind flow is important to eclipse observers, especially those willing to take a chance in the rugged terrain. Winds that flow uphill increase the chances of cloudiness. Those that blow downhill warm and dry the air, dissipating any cloud that may be present.

---

[9] Matveev, V.L. and V.I. Titov, 1985: Data concerning climate structure and variability: global cloudiness fields. Soviet Scientific Research Institute for Hydrometeorological Information - World Data Center.

THE PACIFIC COAST OF PERU AND CHILE

The Peruvian coast line runs nearly parallel to the eclipse track as it approaches from the northwest and the center line runs along the beaches between San Juan and Mollendo. South of Mollendo it leaves the waterfront, turning eastward to climb the steep slopes of the western branch of the Andes, the Cordillera Occidental. The offshore trade winds blow monotonously out of the south, but against the land they are turned by the mountains to flow parallel to the coast. Some of the most persistent sea breezes in the world draw the winds inland and carry cloud against the slopes of the Andes. Day after day a dismal low stratus cloud covers the coast, bringing dense chilly fog and persistent drizzle to the slopes. By November the unyielding cloudiness of winter begins to feel the effects of the upcoming summer and the overcast becomes less tenacious. Occasional sunny days replace the gloom, but the area remains heavily clouded with only a quarter to a third of the mornings showing clear skies. Because the southerly trade winds strike the coast most directly at Tacna, the cloud piles up here more than in any other area.

The secret to finding good eclipse weather along the Peruvian coast is to go inland and uphill. As noted above, the marine cloud is not very deep - usually less than 900 meters, especially in November. Sites can easily be found which remain above the cloud since the land rises very steeply. Satellite pictures show that the cloud hugs the contours of the land, flowing into every coastal valley, and bypassing every ridge and rise of land. One ridge of higher ground that provides a vantage point from which to see the eclipse lies inland and south of Mollendo. Day after day the satellite images showed this ridge protruding above the clouds in the early morning. This ridge and the plateau beyond it is accessible from the Pan American Highway that leads from Arequipa to Mollendo.

If traveling toward Mollendo from Arequipa, don't proceed beyond Guerreros Estación unless the lower slopes are cloud-free. Guerreros Estación is a small railway stop, about 15 kilometers north of the center line, lying at 1140 meters above sea level. This should be high enough to escape all but the deepest marine cloudiness, but it will come with a small time penalty because of its distance from the center of the track. If you insist on trying for the center line at Mollendo, the best strategy is to find the edge of the marine cloud on the Arequipa highway and set up in the dry air just above. You can find this point coming from either direction. Those who are most daring will select a spot at the cloud edge, trusting that the gray mist will not move farther inland as the morning sun warms the slopes. More cautious observers will give up a few seconds of totality to move farther uphill - a few hundred meters at least. Even if Mollendo is clear on eclipse morning, there is a danger that clouds will reform and move inland as temperatures drop ahead of totality. If the air is very humid or clouds have recently cleared, the danger is even greater and a higher altitude is prudent. One clue to watch is the character of the vegetation. Plants near the ocean get their moisture from the fog-drip of the marine clouds. Where clouds are rare, the vegetation will be sparse or of a desert species. Ask around locally.

A more promising location near Arequipa lies at the point where the Pan American Highway crosses the center line on the slopes between Ilo and Moquegua. It is a longer distance to travel but the combination of center line access and good weather prospects make it an attractive alternative. The satellite images from 1991 and 1992 showed that this spot was cloud free about 60% of the time (longitude 71). It is also readily accessible from Tacna. Tacna itself is a cloudy spot, so the trip up the slopes to drier skies may require plenty of lead time. Since the marine cloud may be pushed right up against the mountain side, conditions could be foggy and damp along the road. Tacna has one of the highest frequencies of fog in Table 14. If you travel at night to catch this sunrise eclipse at the center line, be especially generous with your time as visibility may then be at its poorest due to overnight and morning fog. Roads northeast from Tacna lead directly to the center line near Tarata, rather than northwestward to Moquegua. In Figure 9, this spot has a 15% greater frequency of sunny weather (longitude 70°W) than at Moquegua. If the route is passable, this may be the best location from Tacna, especially as it is a much shorter journey.

The statistics at Arequipa, which Table 14 shows to be a very sunny location, are typical for the skies on the slopes above the marine cloud deck. At an altitude of nearly 8500 feet, Peru's second city is well above the coastal cloud. The climatological record shows that sunny skies are measured on two days out of three at eclipse time.

Chile also offers an opportunity to reach the middle of the eclipse path as the shadow track does cut the extreme northeast corner of the country. The road leads northward and upward from Arica, and is apparently very rough. Travel will likely be slow and bumpy, but the route does eventually lead onto the Altiplano and into La Paz. The rail line from coastal Arica to the altiplano also merits some investigation. The center line above Arica enjoys a high frequency of sunshine - perhaps over 80%.

THE BOLIVIAN ALTIPLANO AND THE ANDES MOUNTAINS

Imagine a wide flat plain, 170 kilometers across, 500 kilometers long, surrounded by some of the highest mountains in the world, and lying at 3 to 4 thousand meters above sea level - as high as Mauna Kea. This is the altiplano or Puno of Peru and Bolivia, homeland of the Incas and magnificent mysterious ruins. What a place to watch an eclipse! Although the site has its problems (especially access) the weather is cooperative. Table 14 shows that Uyuni is the sunniest location in the Andes and Charaña ranks only a little behind Arequipa. Charaña lies on the western side of the altiplano and Uyuni on the east. The two straddle the eclipse line so their climatologies should give a reliable indication of the weather across the region. Figure 9 is even more encouraging. During the two years that these data were collected the altiplano was the sunniest location on the eclipse track. The weather prospects in this area are very good for two reasons. First, the altiplano is very high and the air contains less water vapor at 4 thousand meters. Second, the fortress of peaks to the east and west of the Puno wring most of the moisture from any breezes that approach from the tropical interior of the continent or from the Pacific. Winds blow downhill from all directions, drying and dissipating most of the low and middle level clouds that approach. High ice crystal cirrus is the most common cloud, blowing above and off the mountain tops, and often covering a large portion of the sky in a thin veil. Cirrus would not usually hide the eclipse unless it is very thick. Although the frequency of completely clear skies is less than the cloudy coast of Peru, the typical cloud is thin and appears as a transparent cirrus veil or as small patches hanging off mountains in the distance.

When bad weather does invade the plateau, it comes from strong weather systems approaching from Paraguay and Argentina. These systems are high enough to spill over mountain barricades and can fill the altiplano with cloud for a day or two. Some of these low pressure disturbances continue right on to the Pacific, bringing deep layers of cloud and even a little rain to Arequipa and other parts of the Peruvian coast. Because these storms contain plenty of high level cloud, they can be seen approaching from a long distance. Unfortunately, they are also very large systems and can be very difficult to avoid.

Thunderstorms are legendary on the Puno with intense convection forming on nearly every afternoon in the summer. Rain, hail, and strong gusty winds come with the build-ups, often bringing dust and blowing salt from the large salt flats. Fortunately, summer is in its earliest stages in November, and the eclipse track crosses the driest part of the altiplano. Moreover, the thunderstorms are mostly confined to the afternoons and the eclipse is over before 9 AM. Since the source of the moisture that feeds the build-ups comes from the east, sites on the west side of the Puno are more likely to be free of interference. Sucre reports thunderstorms on 2 days of the month on average, but at Charaña it is only once every three years. All-in-all it is difficult to pick the best location on the altiplano. Figure 9 suggests that it might be on the west side near Charaña while Table 14 votes for the east near Uyuni. From the flow of the weather and sources of moisture, the west side appears more promising but in any case the difference is small.

Travel on the altiplano is not difficult though access is not convenient to most of the eclipse track. Potosí is the closest large city to the path, but it lies within the eastern branch of the Andes some distance below the altiplano. To reach the center line on the altiplano is probably easier from Oruro, though the distance is in the neighborhood of 150 kilometers. This is adventure traveling, and the 3800 meter altitude should not be taken lightly. Acclimatization is required. The brain doesn't want to think clearly in an environment with only 60% the oxygen content of sea level. Eclipse observers should plan on several days at altitude beforehand and not a quick visit from sea level. It is probably a good idea to limit the tasks you want to do during the 3 minutes of totality because of the chance of altitude-induced befuddlement.

As the Moon's shadow leaves the altiplano, the weather becomes increasingly cloudy as the path descends to lower pressures on the Atlantic side of the Andes. The increase is gradual at first, so locations around and east of Potosí are quite promising. One good site, where the Pan American Highway crosses the center line south of Potosí, had more than an 80% frequency of sunny skies in 1991 and 1992 according to Figure 9. However this site could have been sunnier than usual over those two years. La Quiaca in Argentina, the closest climatological station, shows a frequency of clear skies (13.7) much lower than Uyuni in Table 14, though still good for the region.

Cloud cover is extremely variable within this region, depending on the exposure of each valley to the east and south winds that bring the moisture. The mountain slopes heat quickly during the morning, drawing in moist air from the lowlands and foothills. These winds are forced to travel uphill, gradually cooling and becoming more humid until cloud begins to form. The clouds grow through the afternoon until deep enough to bring showers and thundershowers.

Cloud-making processes also occur at night. As long as winds are blowing uphill from the east, clouds will form along the Paraguay-Bolivia border. Figure 9 shows this process very well. Longitudes 63°W and 64°W are the cloudiest of the whole track except along the Peruvian coast. In Table 14, Tarija and Yacuiba have one of the lowest frequencies of sunny skies. The area along the Bolivia-Paraguay border should be avoided for this eclipse.

Before leaving Bolivia and the altiplano, it is worthwhile looking at the temperatures. Though November is the warmest month, the altiplano is a very high plain and the mercury can drop sharply overnight. Mean daytime highs at Charaña reach a pleasant 21°C (72°F) but nighttime lows average –5°C (22°F) and the record low is a frigid –15°C (5°F). Travel with warm clothing. Since the eclipse occurs in the early morning on the altiplano, it is entirely possible that temperatures will be below the freezing point. Make sure your equipment and clothing can handle it.

PARAGUAY AND THE GRAN CHACO

The Chaco is a northward extension of the foothill plains of Argentina into western Paraguay and eastern Bolivia. While the western portions are pretty cloudy, sunnier weather returns once the track crosses the Chaco and reaches Concepción. In Figure 9, the frequency of sunshine rebounds to about 70% for the two years studied. This is also reflected in Figure 8, which shows that large scale cloud cover reaches a minimum around Asunción during November. The region is visited by traveling weather disturbances that bring cold fronts from the south into contact with the moist air masses from the northeast. Extensive cloudiness and a catalogue of cloud types comes with these disturbances. These larger weather systems can usually be forecast well in advance (at least as easily as in North America!) and a well-planned eclipse expedition will be able to respond several days ahead. Most large disturbances have thinner spots and holes where mobile observers can gather. This is not an unfailing rule, as some satellite pictures in 1991 and 1992 showed the entire eclipse track covered in cloud from the Chaco to the Atlantic! When the fronts are far to the south, unorganized afternoon cloudiness and thundershowers may build. These usually start well after the appointed time of the shadow passage, leaving the mornings with brighter prospects.

Figure 9 suggests that no particular direction is favored in central and eastern Paraguay and in Argentina. The large scale weather patterns affect them all quite equally. Table 14 is a little more ambiguous since a few sites (Villarrica in particular) claim a very high frequency of clear skies. These conflict with the climatology of the area and are probably due to a very short period of record or erroneous data collection. For a general look at the prospects of the region, the statistics for Asunción are best since it has the longest record of any in the country.

The best strategy to handle weather in Paraguay and Argentina is to obtain forecasts several days in advance and have several widely dispersed sites available. The middle of the track can be reached on good highways by traveling northwest, north and east out of Asunción, and west or south out of Foz do Iguacú (Iguassu Falls). While all three could be cloudy, chances are one will offer better prospects than the others. Remember that weather systems can be very large in this region, and considerable travel might be needed to find sunny skies.

EASTERN PARAGUAY, NORTHERN ARGENTINA, AND BRAZIL

East of Iguassu Falls and the small piece of Argentina that sticks into the eclipse track, the Moon's shadow crosses an extensive area of rolling hills that grow gradually in height to form the 2000 meter high Serra do Mar that guards the Atlantic coast of Brazil. In November, the region is alternately affected by moist unstable air masses from the north and cooler southerlies from higher latitudes. Cold fronts that separate the two air masses can generate large areas of heavy cloud and bring poor eclipse prospects. Movement of fronts is impeded by the mountains and bad weather may linger in the area for several days, spawning a series of small rainy disturbances that keep the skies from clearing. Though they are not high mountains, the Serra also provide some blocking of moist airflows from the Atlantic, preventing them from moving into Argentina and Paraguay. Figure 9 shows this as a small decrease in sunny skies near Criciúma and along the Atlantic coast. In particular, the coast itself at 49°W is cloudiest of all. The statistics in Table 14 reflect this climatology as well. For instance, compare Posadas and Alegrete with Florianapolis and Porto Alegre.

It is a good rule to avoid mountains during an eclipse, particularly those in which a nearby ocean is available to supply moisture. Valleys have an unfortunate tendency to fill with cloud as second contact approaches and it is difficult to find a site with reliable downslope drying winds in the jumble of terrain. The eclipse track through Brazil is just such an area and the statistics bear out its poor ranking. It has one tempting asset - a four minute eclipse. It is advisable to use weather forecasts a day or two ahead of time to plan your location. However, a long trip may be necessary to find better weather, perhaps into Argentina or Paraguay.

OFFSHORE

Cloudiness along the coast of Brazil is unusually high because of the blocking influence of the Serra do Mar, but the effect does not extend more than a hundred kilometers or so offshore. Conditions improve farther out to sea. Figure 8 shows that the area along the track is quite promising for several hundred kilometers beyond the coast. While the Atlantic coast of Brazil was very close to the edge of the satellite images examined to develop the statistics in Figure 9, it appeared that offshore weather systems were spotted with numerous breaks and holes. These holes and the patchy nature of the disturbances are probably due to the influence of the anticyclone which resides off the African coast. Ships have excellent mobility with which to exploit these breaks, and a water-borne eclipse chase should be quite promising. Weather forecasts and satellite images can provide enough information to select a promising location on the track, but leave lots of time to reach it in case the distance to clear skies is large.

Wave heights are a major concern for ship-board eclipse observers since they make photography difficult (and interfere with lunch!). Mean wave heights off Brazil range between 1 and 1.5 meters in November, a low value for the latitude. This is comparable to wave heights in waters near the Philippines for those who caught the 1988 eclipse from the ocean surface. The standard deviation of the wave height is close to one meter along the east coast of South America. This means there is a 66% chance that wave heights will lie between one-half and 2.5 meters on eclipse day. Of course the exact value on eclipse day will depend on prevailing winds and the location of nearby storms.

Wave heights, cloudiness and dismal prospects for the eclipse all increase along the eclipse track as it moves onto the African side of the South Atlantic. Gough Island off of South Africa has a high mean cloudiness and few sunshine hours for the month. Once around Cape Horn however, brighter weather begins to return, though Figure 8 is only marginally encouraging.

SUMMARY

This eclipse offers weather and adventure to suit all tastes. The better weather prospects are on the west side of South America, either near Arequipa or on the altiplano. Prospects along the eastern Andes are poor but promising through Paraguay and Argentina. The Pacific and Atlantic coasts offer the poorest prospects.

OBSERVING THE ECLIPSE

EYE SAFETY DURING SOLAR ECLIPSES

The Sun can be viewed safely with the naked eye only during the few brief minutes of a *total* solar eclipse. Partial and annular solar eclipses are *never* safe to watch without taking special precautions. Even when 99% of the Sun's surface is obscured during the partial phases, the remaining photospheric crescent is intensely bright and cannot be viewed directly without eye protection [Chou, 1981; Marsh, 1982]. *Do not attempt to observe the partial or annular phases of any eclipse with the naked eye. Failure to use appropriate filtration may result in permanent eye damage or blindness!*

Generally, the same equipment, techniques and precautions used to observe the Sun outside of eclipse are required [Pasachoff & Menzel, 1992; Sherrod, 1981]. There are several safe methods which may be used to watch the partial phases. The safest of these is projection, in which a pinhole or small opening is used to cast the image of the sun on a screen placed a half-meter or more beyond the opening. Projected images of the sun may even be seen on the ground in the small openings created by interlacing fingers, or in the dappled sunlight beneath a tree. Binoculars can also be used to project a magnified image of the sun on a white card, but you must avoid the temptation of using these instruments for direct viewing.

Direct viewing of the sun should only be done using filters specifically designed for this purpose. Such filters usually have a thin layer of aluminum, chromium or silver deposited on their surfaces which attenuates both the visible and the infrared energy. Experienced amateur and professional astronomers may use one or two layers of completely exposed and fully developed black-and-white film, provided the film contains a silver emulsion. Since developed color films lack silver, they are unsafe for use in solar viewing. A widely available alternative for safe eclipse viewing is a number 14 welder's glass. However, only mylar or glass filters specifically designed for the purpose should used with telescopes or binoculars.

Unsafe filters include color film, smoked glass, photographic neutral density filters and polarizing filters. Deep green or gray filters often sold with inexpensive telescopes are also dangerous. They should not be used for viewing the sun at any time since they often crack from overheating. Do not experiment with other filters unless you are certain that they are safe. Damage to the eyes comes predominantly from invisible infrared wavelengths. The fact that the sun appears dark in a filter or that you feel no discomfort does not guarantee that your eyes are safe. Avoid all unnecessary risks. Your local planetarium or amateur astronomy club is a good source for additional information.

SKY AT TOTALITY

The total phase of an eclipse is accompanied by the onset of a rapidly darkening sky whose appearance approximates that of evening twilight 30 to 45 minutes after sunset. The effect presents an excellent opportunity to view planets and bright stars in the daytime sky. Such observations are useful in gauging the apparent sky brightness and transparency during totality. The Sun is in Libra and a number of planets and bright stars will be above the horizon for observers within the umbral path. Figures 10 and 11 depict the appearance of the sky from the western and eastern sections, respectively, of the South American path. Venus is usually the brightest planet and can actually be observed in broad daylight provided that the sky is cloud free and of high transparency (i.e. - no dust or particulates). During the Nov 1994 eclipse, Venus is only half a day past inferior conjunction and will be located a mere 5° west of the Sun. Look for the planet during the partial phases by first covering the eclipsed Sun with an extended hand. During totality, it will be almost impossible to miss Venus since it is so close to the Sun and will shine at a magnitude of $m_v=-4.0$. Although two magnitudes fainter, Jupiter will also be well placed 11° east of the Sun and shining at $m_v=-1.7$. Under good conditions, it may be possible to spot Jupiter 5 to 10 minutes before totality. Mercury is approaching eastern elongation on 6 Nov and is located 18° west of the Sun at $m_v=-0.1$. Although a bit more challenging, it should still be easy to see provided skies are clear. Spica ($m_v=+0.7$) is 4° south of Mercury which may be used as a guide to locate it. The most difficult of the naked eye planets will be Mars ($m_v=+0.7$). It is located 85° west of the Sun, and 14° west of Regulus ($m_v=+1.35$). Saturn is 115° east of the Sun and will be below the horizon for all observers in South America. Other stars to look for include Antares ($m_v=+0.9v$), Arcturus ($m_v=-0.04$), Alpha and Beta Cen ($m_v=-0.01$ & $m_v=+0.6v$), Canopus ($m_v=-0.72$), Sirius ($m_v=-1.46$) and Procyon ($m_v=+0.38$).

The following ephemeris [using Bretagnon and Simon, 1986] gives the positions of the naked eye planets during the eclipse. **Delta** is the distance of the planet from Earth (A.U.'s), **V** is the apparent visual magnitude of the planet, and **Elong** gives the solar elongation or angle between the Sun and planet. Note that Jupiter is near opposition and will be below the horizon during the eclipse for all observers.

```
Planetary Ephemeris:    3 Nov 1994  14:00:00 UT    Equinox = Mean Date

Planet        RA           Dec         Delta    V   Diameter  Phase   Elong
           h   m   s      °   '   "                     "               °
Sun        14  33  59   -15-06-07    0.99193  -26.7  1934.9    -        -
Mercury    13  26  23   -06-48-50    0.92429   -0.1     7.3   0.47    18.5W
Venus      14  23   2   -19-46-46    0.27023   -4.0    61.8   0.00     5.4W
Mars       09  14  28    17 42 34    1.32296    0.7     7.1   0.89    85.3W
Jupiter    15  19  36   -17-31-09    6.36421   -1.7    30.9   1.00    11.2E
Saturn     22  32  52   -11-12-28    9.24022    0.2    17.9   1.00   114.8E
```

ECLIPSE PHOTOGRAPHY

The eclipse may be safely photographed provided that the above precautions are followed. Almost any kind of camera with manual controls can be used to capture this rare event. However, a lens with a fairly long focal length is recommended to produce as large an image of the Sun as possible. A standard 50 mm lens yields a minuscule 0.5 mm image, while a 200 mm telephoto or zoom produces a 1.9 mm image. A better choice would be one of the small, compact catadioptic or mirror lenses which have become widely available in the past ten years. The focal length of 500 mm is most common among such mirror lenses and yields a solar image of 4.6 mm. Adding 2x tele-converter will produce a 1000 mm focal length which doubles the Sun's size to 9.2 mm. Focal lengths in excess of 1000 mm usually fall within the realm of amateur telescopes. If full disk photography of partial phases on 35 mm format is planned, the focal length of the telescope or lens must be 2600 mm or less. Longer focal lengths will only permit photography of a portion of the Sun's disk. Furthermore, in order to photograph the Sun's corona during totality, the focal length should be no longer than 1500 mm to 1800 mm (for 35 mm equipment). For any particular focal length, the diameter of the Sun's image is approximately equal to the focal length divided by 109.

A mylar or glass solar filter must be used on the lens at all times for both photography and safe viewing. Such filters are most easily obtained through manufacturers and dealers listed in *Sky & Telescope* and *Astronomy* magazines. These filters typically attenuate the Sun's visible and infrared energy by a factor of 100,000. However, the actual filter attenuation and choice of ISO film speed will play critical roles in determining the correct photographic exposure. A low to medium speed film is recommended (ISO 50 to 100) since the Sun gives off abundant light. The easiest method for determining the correct exposure is accomplished by running a calibration test on the uneclipsed Sun. Shoot a roll of film of the mid-day Sun at a fixed aperture [f/8 to f/16] using every shutter speed between 1/1000 and 1/4 second. After the film is developed, the best exposures are noted and may be used to photograph all the partial phases since the Sun's surface brightness remains constant throughout the eclipse.

Certainly the most spectacular and awe inspiring phase of the eclipse is totality. For a few brief minutes, the Sun's pearly white corona, red prominences and chromosphere are visible. The great challenge is to obtain a set of photographs which capture some aspect of these fleeting phenomena. The most important point to remember is that during the total phase, all solar filters *must be removed!* The corona has a surface brightness a million times fainter than the photosphere, so photographs of the corona are made without a filter. Furthermore, it is completely safe to view the totally eclipsed Sun directly with the naked eye. No filters are needed and they will only hinder your view. The average brightness of the corona varies inversely with the distance from the Sun's limb. The inner corona is far brighter than the outer corona. Thus, no one exposure can capture its the full dynamic range. The best strategy is to choose one aperture or f/number and bracket the exposures over a range of shutter speeds (i.e. - 1/1000 down to 1 second). Rehearsing this sequence is highly recommended since great excitement accompanies totality and there is little time to think.

Exposure times for various combinations of film speeds (ISO), apertures (f/number) and solar features (chromosphere, prominences, inner, middle and outer corona) are summarized in Table 15. To use the table, first select the ISO film speed in the upper left column. Now, move to the right to the desired aperture or f/number for the chosen ISO. The shutter speeds in that column may be used as starting points for photographing various features and phenomena tabulated in the 'Subject' column at the far left. For example, to photograph prominences using ISO 100 at f/11, the table recommends an exposure of 1/500. Alternatively, you can calculate the recommended shutter speed using the 'Q' factors tabulated along with the exposure formula at the bottom of Table 15. Keep in mind that these exposures are based on a clear sky and an average corona. You should bracket your exposures to take into account the actual sky conditions and the variable nature of these phenomena.

Another interesting way to photograph the eclipse is to record its various phases all on one frame. This is accomplished by using a stationary camera capable of making multiple exposures (check the camera instruction manual). Since the Sun moves through the sky at the rate of 15 degrees per hour, it slowly drifts through the field of view of any camera equipped with a normal focal length lens (i.e. - 35 to 50 mm). If the camera is oriented so that the Sun drifts along the frame's diagonal, it will take over three hours for the Sun to cross the field of a 50 mm lens. The proper camera orientation can be determined through trial and error several days before the eclipse. This will also insure that no trees or buildings obscure the camera's view during the eclipse. The Sun should be positioned along the eastern (left) edge or corner of the viewfinder shortly before the eclipse begins. Exposures are then made throughout the eclipse at five minute intervals. The camera must remain perfectly rigid during this period and may be clamped to a wall or fence post since tripods are easily bumped. The final photograph will consist of a string of Suns, each showing a different phase of the eclipse.

Finally, an eclipse effect which is easily captured with point-and-shoot or automatic cameras should not be overlooked. During the eclipse, the ground under nearby shade trees is covered with small images of the crescent Sun. The gaps between the tree leaves act like pinhole cameras and each one projects its own tiny image of the Sun. The effect can be duplicated by forming a small aperture with one's hands and watching the ground below. The pinhole camera effect becomes more prominent with increasing eclipse magnitude. Virtually any camera can be used to photograph the phenomenon, but automatic cameras must have their flashes turned off since this will obliterate the pinhole images.

For more information on eclipse photography, observations and eye safety, see FURTHER READING in the BIBLIOGRAPHY.

CONTACT TIMINGS FROM THE PATH LIMITS

Precise timings of second and third contacts, made near the northern and southern limits of the umbral path (i.e. - the graze zones), are of value in determining the diameter of the Sun relative to the Moon at the time of the eclipse. Such measurements are essential to an ongoing project to monitor changes in the solar diameter. Due to the conspicuous nature of the eclipse phenomena and their strong dependence on geographical location, scientifically useful observations can be made with relatively modest equipment. Inexperienced observers are cautioned to use great care in making such observations. The safest timing technique consists of the inspection of a projected image of the rather than direct viewing of the solar disk. The observer's geodetic coordinates are required and can be measured from USGS or other large scale maps. If a map is unavailable, then a detailed description of the observing site should be included which provides information such as distance and directions of the nearest towns/settlements, nearby landmarks, identifiable buildings and road intersections. Alternatively, beacon devices are commercially available (~$500 US) which provide the user's location via global positioning satellites. The method of contact timing should also be described, along with an estimate of the error. The precisional requirements of these observations are ±0.5 seconds in time, 1" (~30 meters) in latitude and longitude, and ±20 meters (~60 feet) in elevation. The International Occultation Timing Association (IOTA) coordinates observers world-wide during each eclipse. For more information, contact:

> Dr. David W. Dunham/IOTA
> 7006 Megan Lane
> Greenbelt, MD 20770-3012
> U. S. A.

Send reports containing graze observations, eclipse contact and Baily's bead timings, including those made anywhere near or in the path of totality or annularity to:

> Dr. Alan D. Fiala
> Orbital Mechanics Dept.
> U. S. Naval Observatory
> 3450 Massachusetts Ave., NW
> Washington, DC 20392-5420

PLOTTING THE PATH ON MAPS

If high resolution maps of the umbral path are needed, the coordinates listed in Table 7 are conveniently provided at 1° increments of longitude to assist plotting by hand. The path coordinates in Table 3 define a line of maximum eclipse at five minute increments in Universal Time. It is also advisable to include lunar limb corrections to the northern and southern limits listed in Table 6, especially if observations are planned from the graze zones. Global Navigation Charts (1:5,000,000), Operational Navigation Charts (scale 1:1,000,000) and Tactical Pilotage Charts (1:500,000) of many parts of the world can be obtained from the Defense Mapping Agency. For specific information about map availability, purchase prices, and ordering instructions, call DMA at 1-800-826-0342 (USA) or (301) 227-2495 (outside USA). The address is:

Defense Mapping Agency CSC
Attn: PMA
Washington, DC 20315-0010, USA.

It is also advisable to check the telephone directory for any map specialty stores in your city or metropolitan area. They often have large inventories of many maps available for immediate delivery.

ALGORITHMS, EPHEMERIDES AND PARAMETERS

Algorithms for the eclipse predictions were developed by Espenak primarily from the *Explanatory Supplement* [1974] with additional algorithms from Meeus, Grosjean and Vanderleen [1966]. The solar and lunar ephemerides were generated from the JPL DE200 and LE200, respectively. All eclipse calculations were made using a value for the Moon's radius of k=0.2722810 for umbral contacts, and k=0.2725076 [adopted IAU value] for penumbral contacts. Center of mass coordinates were used except where noted. An extrapolated value for ΔT of 59.5 seconds was used to convert the predictions from Terrestrial Dynamical Time to Universal Time.

The primary source for geographic coordinates used in the local circumstances tables is *The New International Atlas* (Rand McNally, 1991). Elevations for major cities were taken from *Climates of the World* (U. S. Dept. of Commerce, 1972).

ACKNOWLEDGMENTS

Most of the predictions presented in this publication were generated on a Macintosh IIfx. Additional computations, particularly those dealing with Watts' datum and the lunar limb profile were performed on a DEC VAX 11/785 computer. Word processing and page layout for the publication were done on a Macintosh using Microsoft Word v5.1. Figure annotation was done with Claris MacDraw Pro.

We thank Francis Reddy who helped develop the data base of geographic coordinates for major cities used in the local circumstances predictions. Dr. Wayne Warren graciously provided a draft copy of the *IOTA Observer's Manual* for use in describing contact timings near the path limits. We also want to thank Dr. John Bangert for several valuable discussions and for sharing the USNO mailing list for the eclipse *Circulars*. The format and content or this work has drawn heavily upon over 40 years of eclipse *Circulars* published by the U. S. Naval Observatory. We owe a debt of gratitude to past and present staff of that institution who have performed this service for so many years. In particular, we would like to recognize the work of Julena S. Duncombe, Alan D. Fiala, Marie R. Lukac, John A. Bangert and William T. Harris. Dr. Jay M. Pasachoff kindly reviewed the manuscript and offered a number of valuable suggestions. The support of Environment Canada is acknowledged in the acquisition and arrangement of the weather data. Finally, the authors thank Goddard's Laboratory for Extraterrestrial Physics for several minutes of CPU time on the LEPVX2 computer.

The names and spellings of countries, cities and other geopolitical regions are not authoritative, nor do they imply any official recognition in status. Corrections to names, geographic coordinates and elevations are actively solicited in order to update the data base for future eclipses. All calculations, diagrams and opinions presented in this publication are those of the authors and they assume full responsibility for their accuracy.

BIBLIOGRAPHY

REFERENCES

Bretagnon, P. and Simon, J. L., *Planetary Programs and Tables from –4000 to +2800*, Willmann-Bell, Richmond, Virginia, 1986.
Chou, B. R., "Safe Solar Filters," *Sky & Telescope*, August 1981, p. 119.
Climates of the World, U. S. Dept. of Commerce, Washington DC, 1972.
Dunham, J. B, Dunham, D. W. and Warren, W. H., *IOTA Observer's Manual*, (draft copy), 1992.
Espenak, F., *Fifty Year Canon of Solar Eclipses: 1986 - 2035*, NASA RP-1178, Greenbelt, MD, 1987.
Explanatory Supplement to the Astronomical Ephemeris and the American Ephemeris and Nautical Almanac, Her Majesty's Nautical Almanac Office, London, 1974.
Herald, D., "Correcting Predictions of Solar Eclipse Contact Times for the Effects of Lunar Limb Irregularities," *J. Brit. Ast. Assoc.*, 1983, **93**, 6.
Marsh, J. C. D., "Observing the Sun in Safety," *J. Brit. Ast. Assoc.*, 1982, **92**, 6.
Meeus, J., Grosjean, C., and Vanderleen, W., *Canon of Solar Eclipses*, Pergamon Press, New York, 1966.
Morrison, L. V., "Analysis of lunar occultations in the years 1943-1974...," *Astr. J.*, 1979, **75**, 744.
Morrison, L.V. and Appleby, G.M., "Analysis of lunar occultations - III. Systematic corrections to Watts' limb-profiles for the Moon," *Mon. Not. R. Astron. Soc.*, 1981, **196**, 1013.
The New International Atlas, Rand McNally, Chicago/New York/San Francisco, 1991.
van den Bergh, *Periodicity and Variation of Solar (and Lunar) Eclipses*, Tjeenk Willink, Haarlem, Netherlands, 1955.
Watts, C. B., "The Marginal Zone of the Moon," *Astron. Papers Amer. Ephem.*, 1963, **17**, 1-951.

FURTHER READING

Allen, D. and Allen, C., *Eclipse*, Allen & Unwin, Sydney, 1987.
Astrophotography Basics, Kodak Customer Service Pamphlet P150, Eastman Kodak, Rochester, 1988.
Brewer, B., *Eclipse*, Earth View, Seattle, 1991.
Covington, M., *Astrophotography for the Amateur*, Cambridge University Press, Cambridge, 1988.
Espenak, F., "Total Eclipse of the Sun," *Petersen's PhotoGraphic*, June 1991, p. 32.
Fiala, A. D., DeYoung, J. A. and Lukac, M. R., *Solar Eclipses, 1991-2000*, USNO Circular No. 170, U. S. Naval Observatory, Washington, DC, 1986.
Littmann, M. and Willcox, K., *Totality, Eclipses of the Sun*, University of Hawaii Press, Honolulu, 1991.
Lowenthal, J., *The Hidden Sun: Solar Eclipses and Astrophotography*, Avon, New York, 1984.
Mucke, H. and Meeus, J., *Canon of Solar Eclipses: –2003 to +2526*, Astronomisches Büro, Vienna, 1983.
North, G., *Advanced Amateur Astronomy*, Edinburgh University Press, 1991.
Oppolzer, T. R. von, *Canon of Eclipses*, Dover Publications, New York, 1962.
Pasachoff, J. M. and Covington, M., *Cambridge Guide to Eclipse Photography*, Cambridge University Press, Cambridge and New York, 1993.
Pasachoff, J. M. and Menzel, D. H., *Field Guide to the Stars and Planets*, 3rd edition, Houghton Mifflin, Boston, 1992.
Sherrod, P. C., *A Complete Manual of Amateur Astronomy*, Prentice-Hall, 1981.
Sweetsir, R. and Reynolds, M., *Observe: Eclipses*, Astronomical League, Washington, DC, 1979.
Zirker, J. B., *Total Eclipses of the Sun*, Van Nostrand Reinhold, New York, 1984.

TOTAL SOLAR ECLIPSE OF 3 NOVEMBER 1994

FIGURES

Figure 1: ORTHOGRAPHIC PROJECTION MAP OF THE ECLIPSE PATH
Total Solar Eclipse of 3 Nov 1994

Geocentric Conjunction = 13:47:07.9 UT   J.D. = 2449660.074397
Greatest Eclipse = 13:39:06.3 UT   J.D. = 2449660.068823

Eclipse Magnitude = 1.05351   Gamma = -0.35216   ΔT = 59.7 s

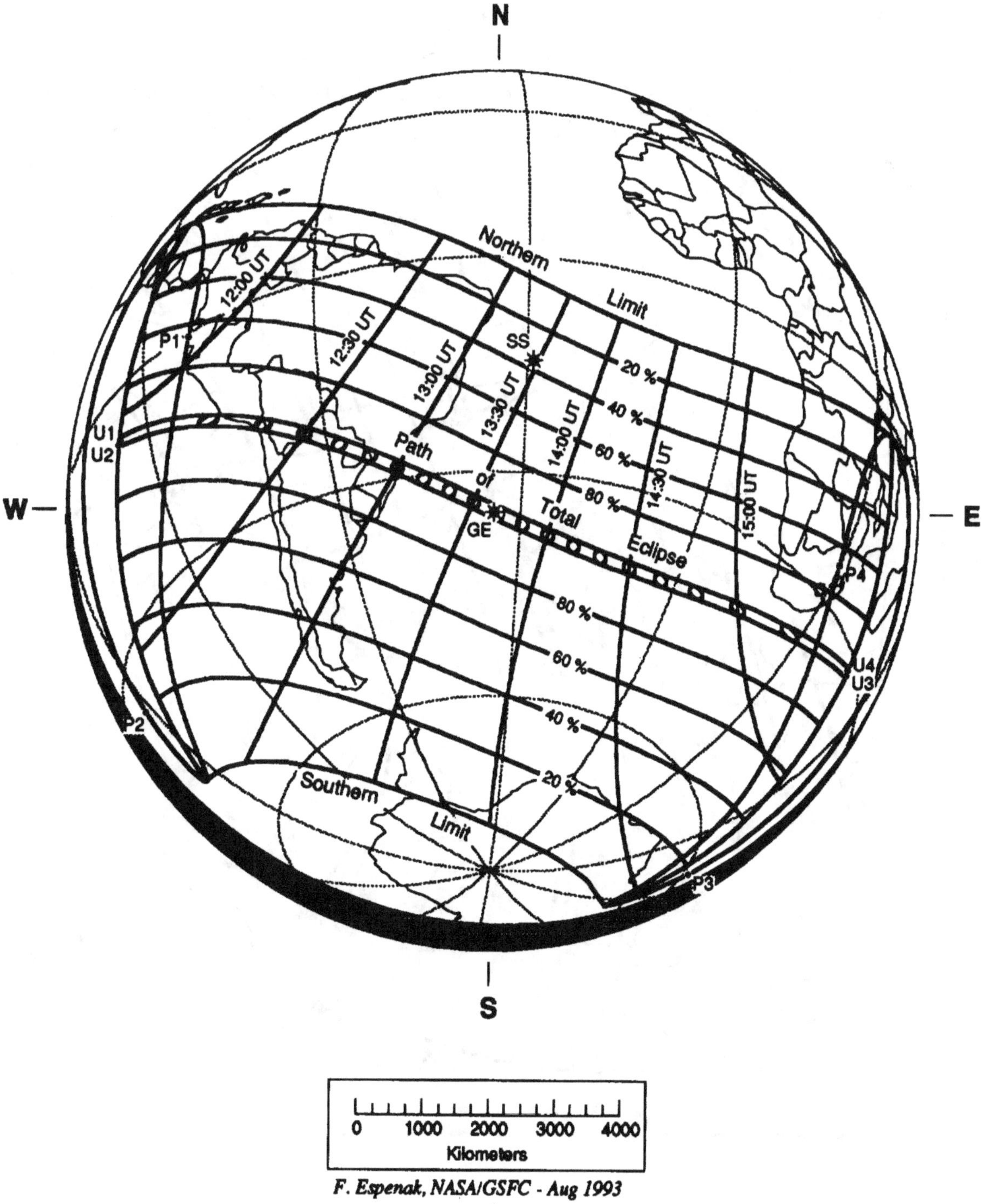

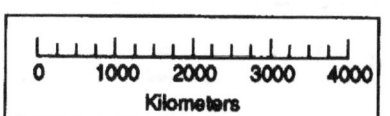

F. Espenak, NASA/GSFC - Aug 1993

Figure 2: Stereographic Projection Map of The Eclipse Path
Total Solar Eclipse of 3 Nov 1994

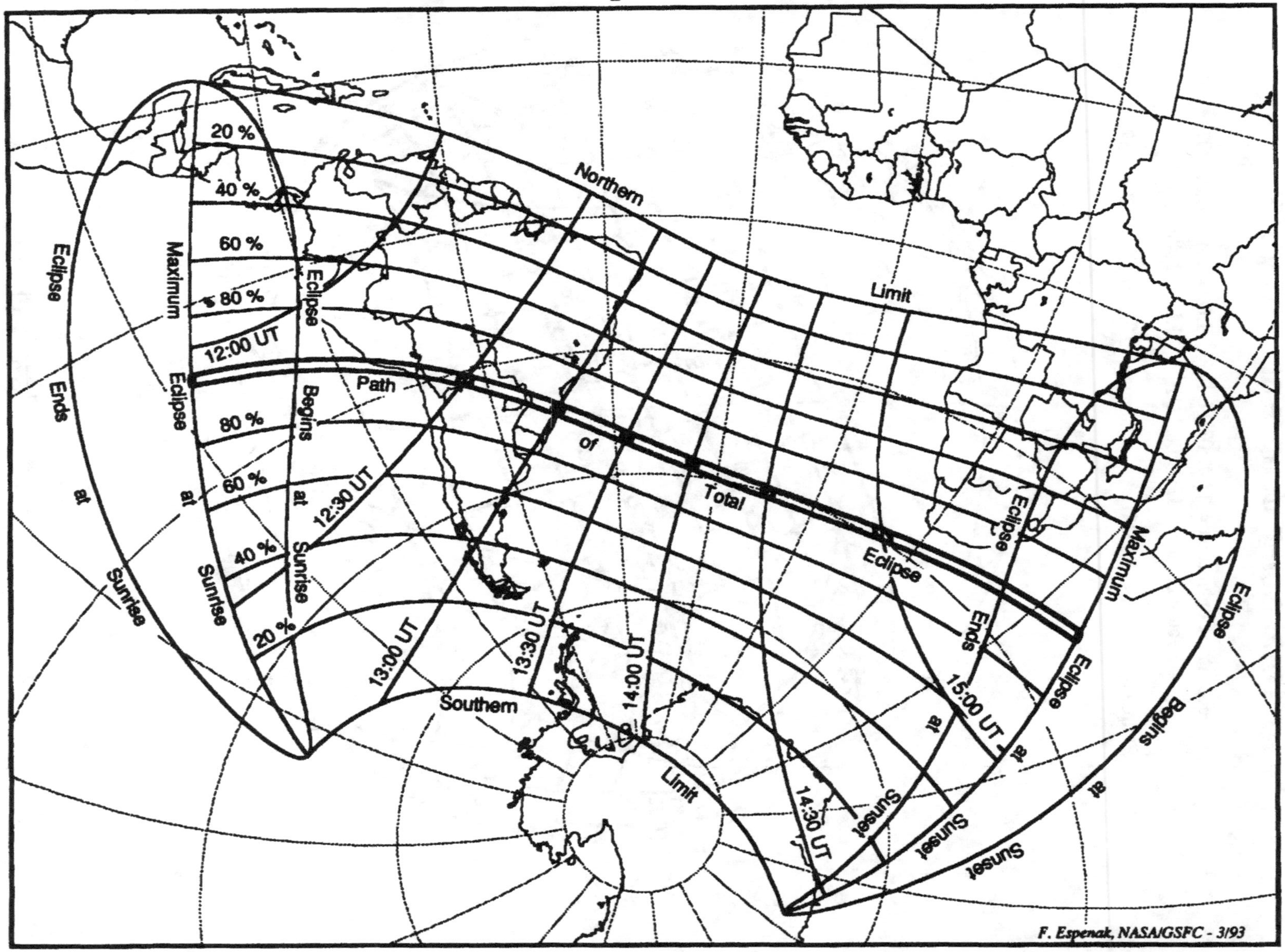

Figure 3: **The Eclipse Path in South America**

Total Solar Eclipse of 3 Nov 1994

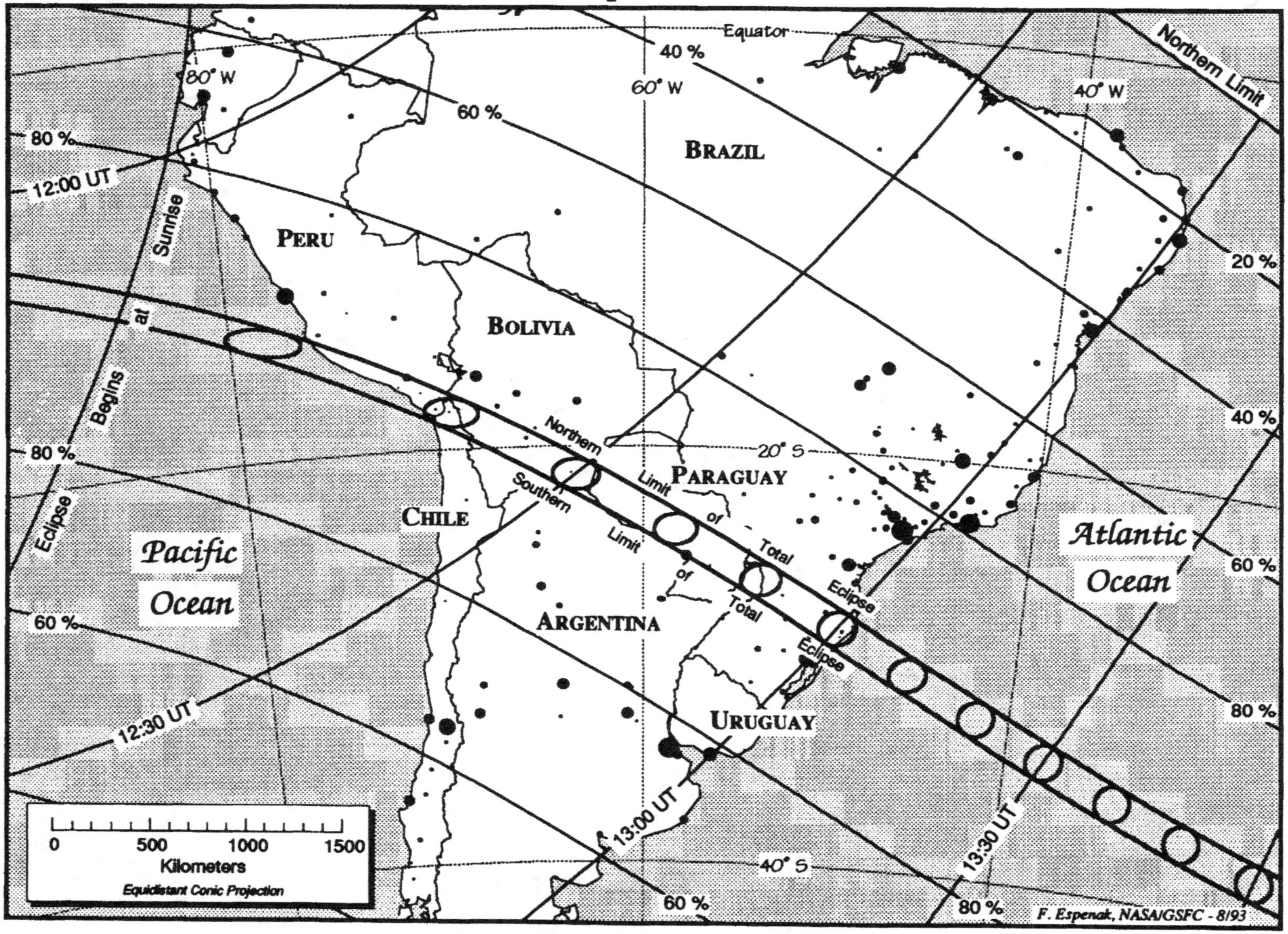

Figure 4: The Eclipse Path in Western South America
Total Solar Eclipse of 3 Nov 1994

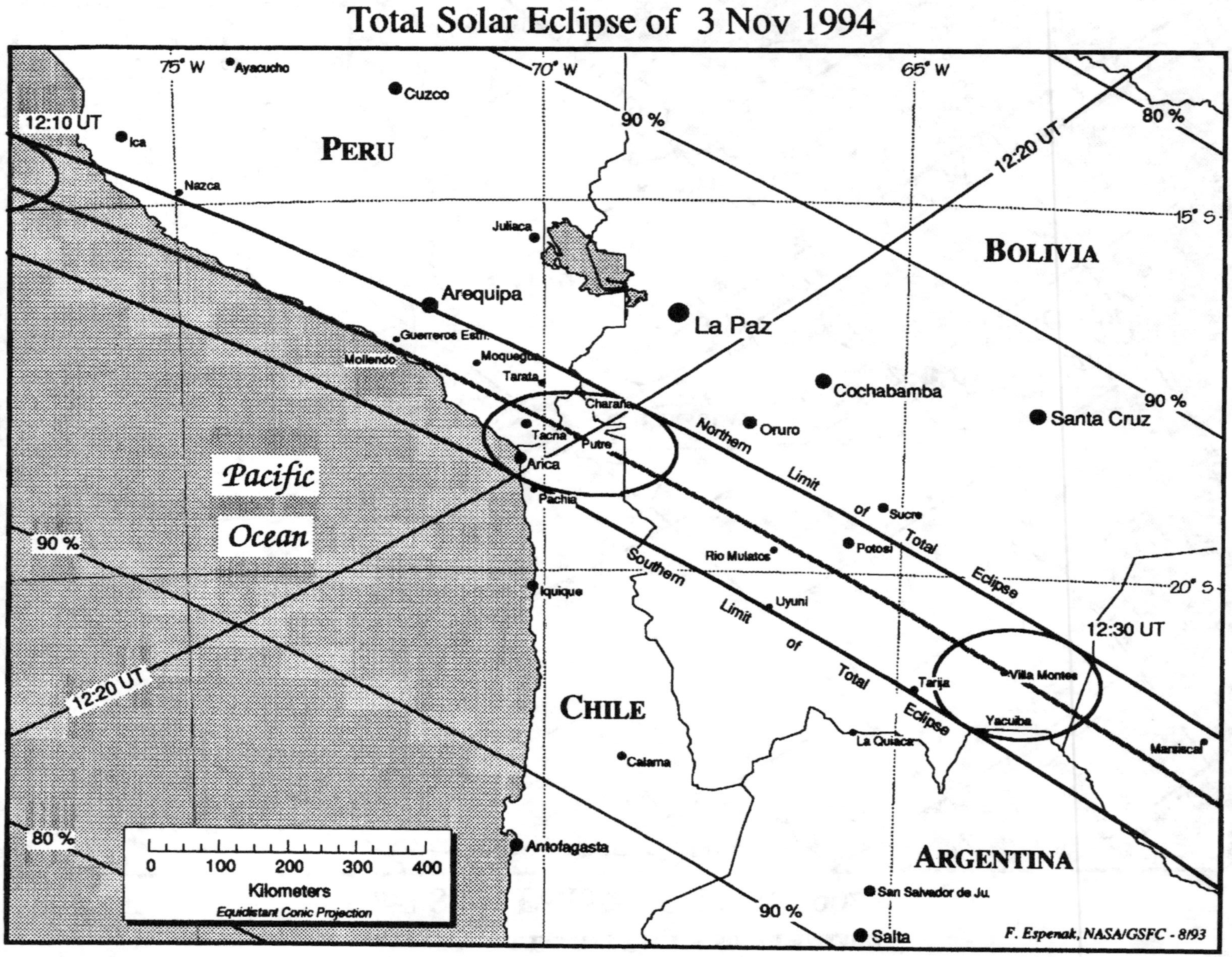

Figure 5: **THE ECLIPSE PATH IN EASTERN SOUTH AMERICA**
Total Solar Eclipse of 3 Nov 1994

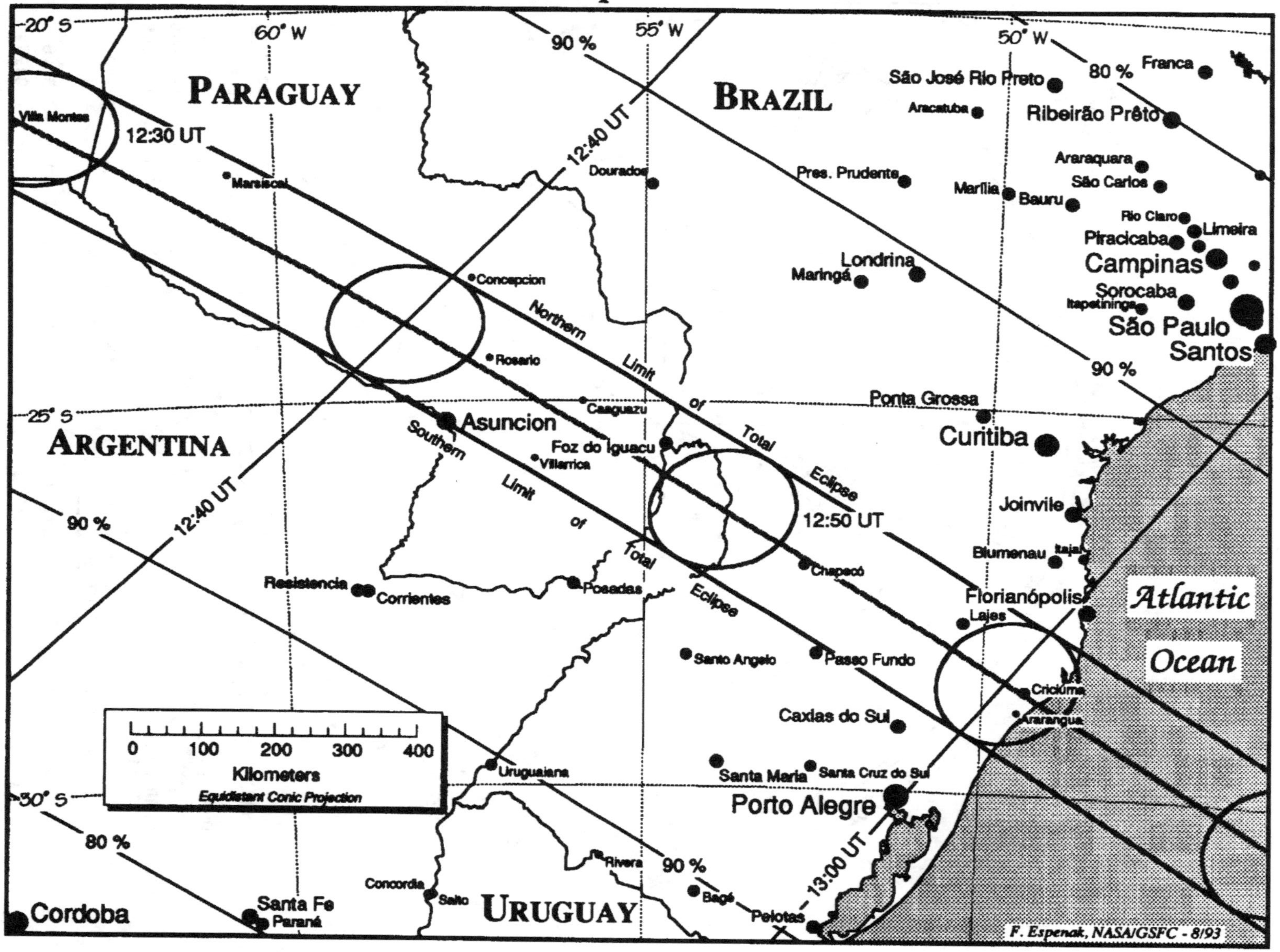

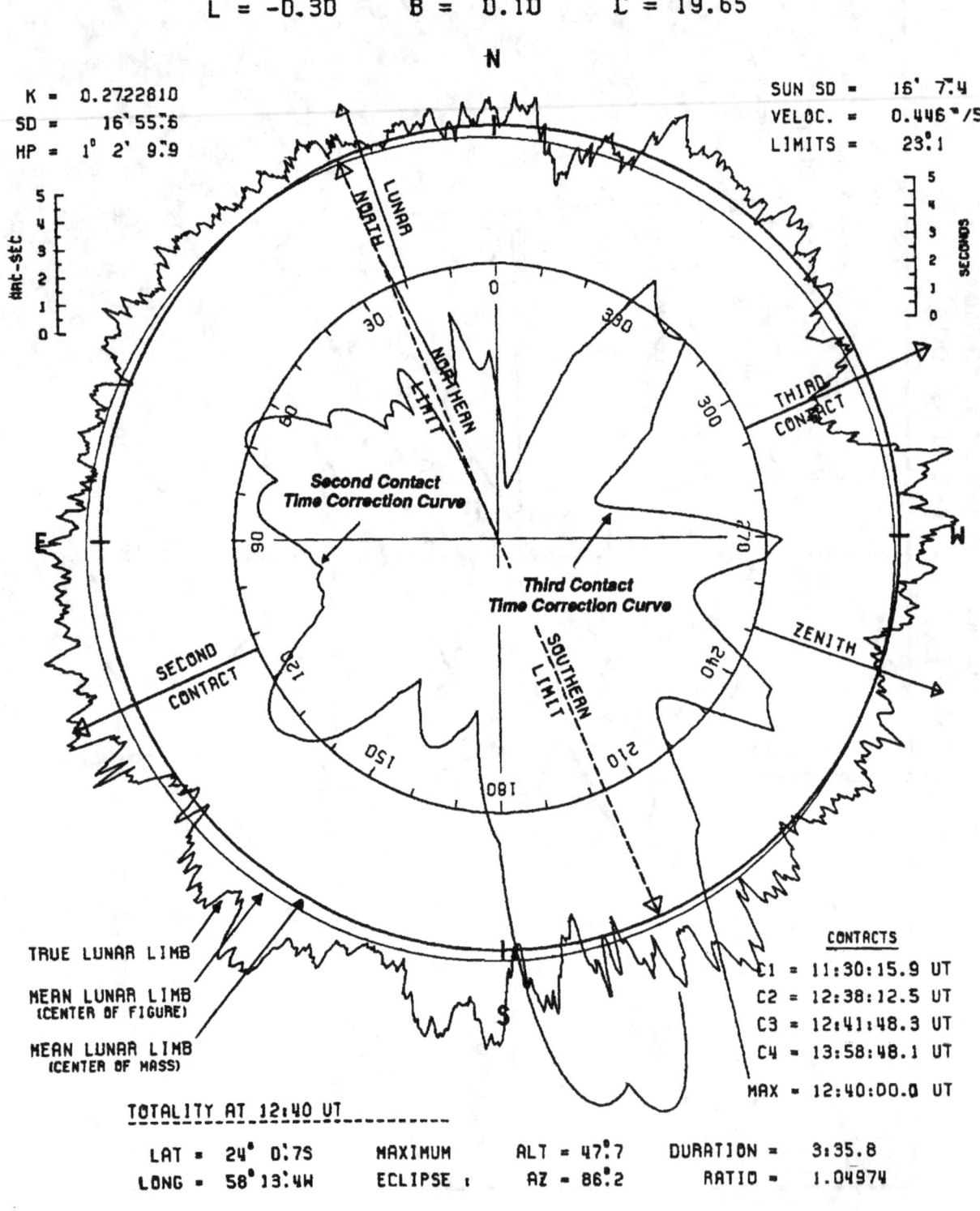

Figure 6: THE LUNAR LIMB PROFILE AT 12:40 UT
Total Solar Eclipse of 3 Nov 1994

Figure 7: **Mean Surface Pressure for November**

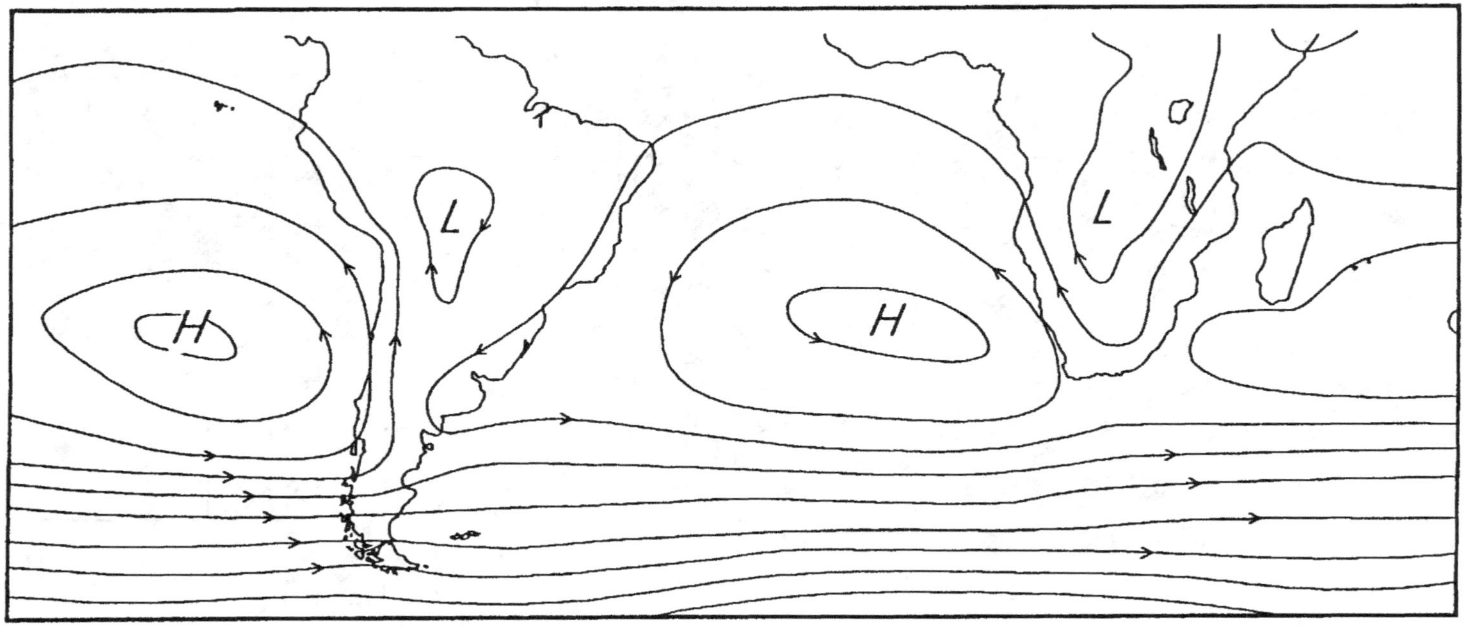

Figure 8: **Mean Cloud Cover for November**

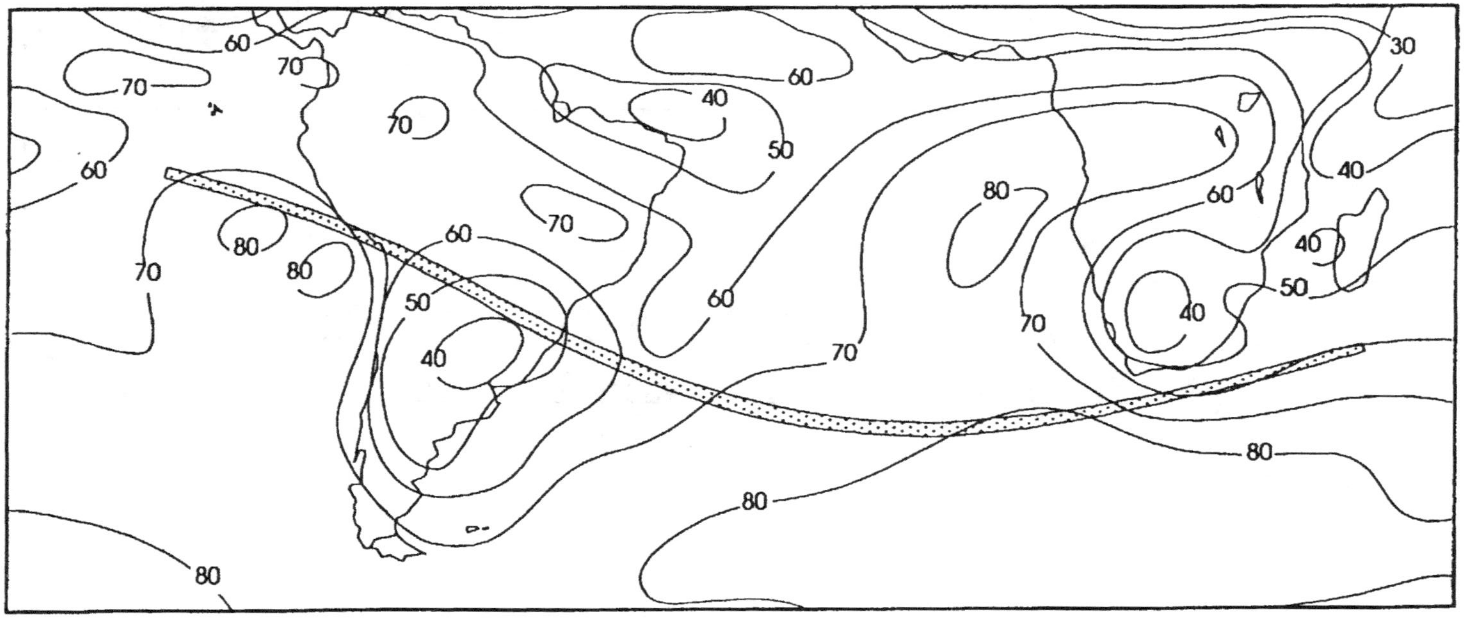

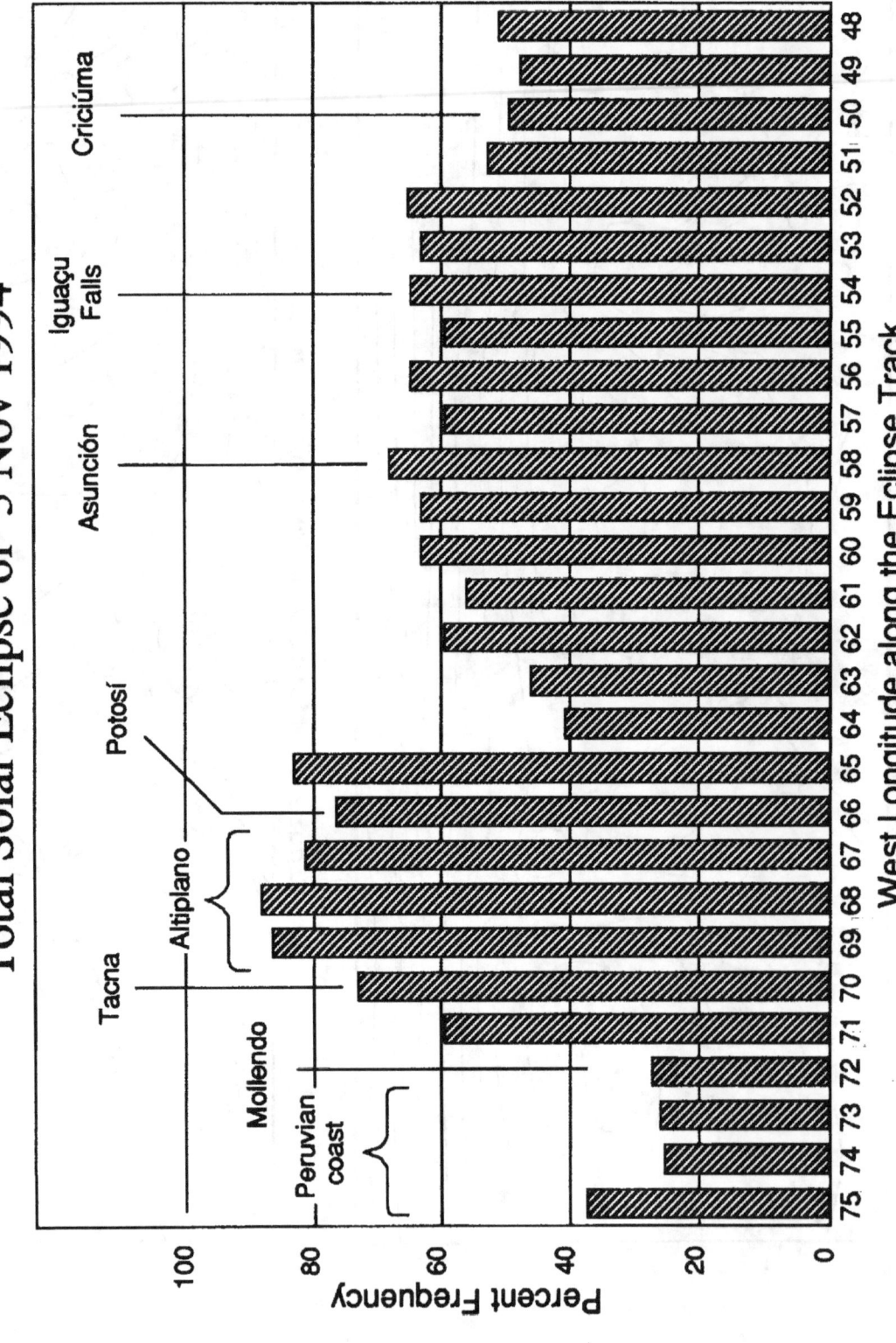

Figure 9: FREQUENCY OF CLEAR SKIES ALONG THE ECLIPSE PATH
Total Solar Eclipse of 3 Nov 1994

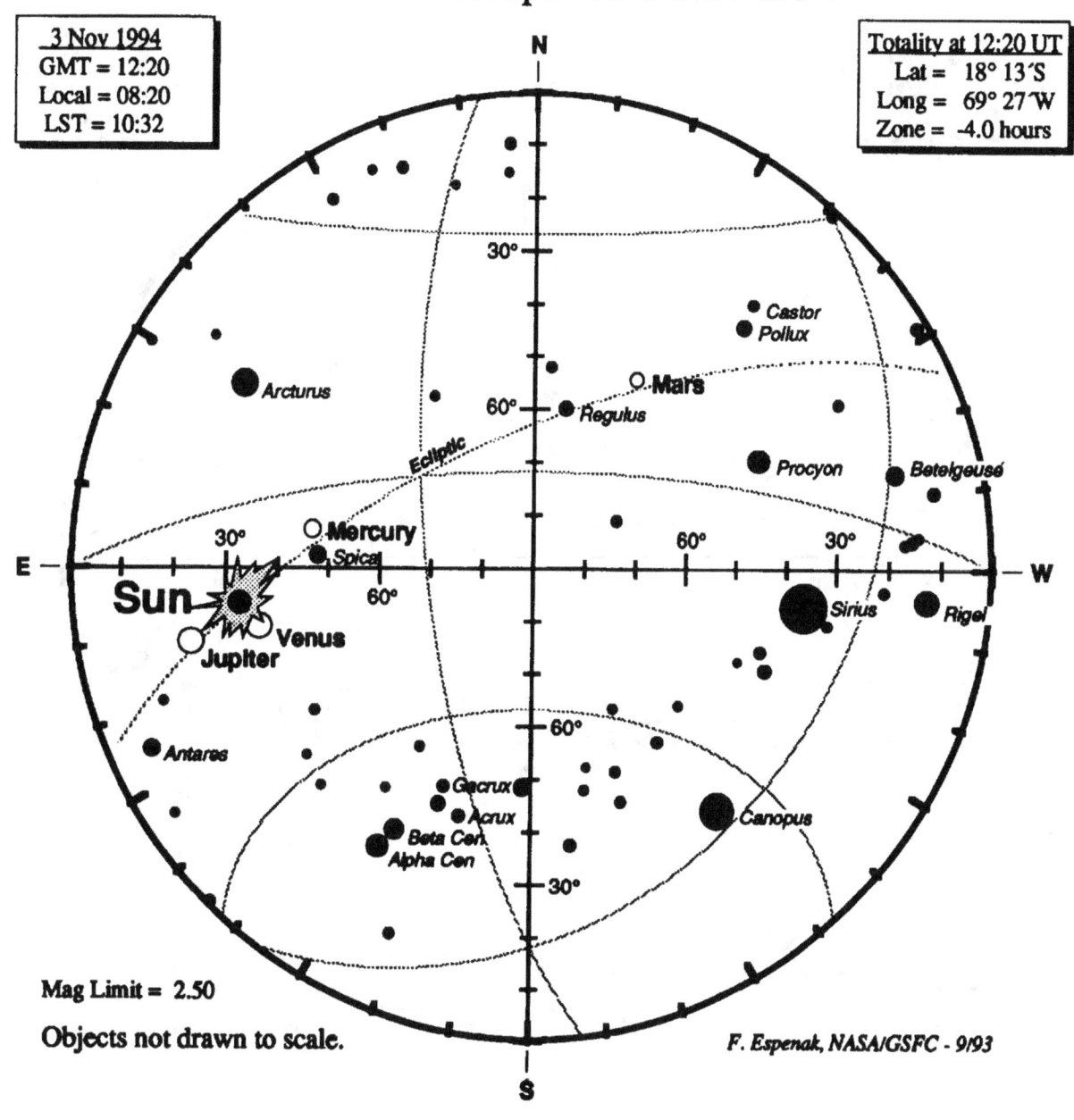

Figure 10: THE SKY DURING TOTALITY AS SEEN AT 12:20 UT

Figure 11: THE SKY DURING TOTALITY AS SEEN AT 13:00 UT

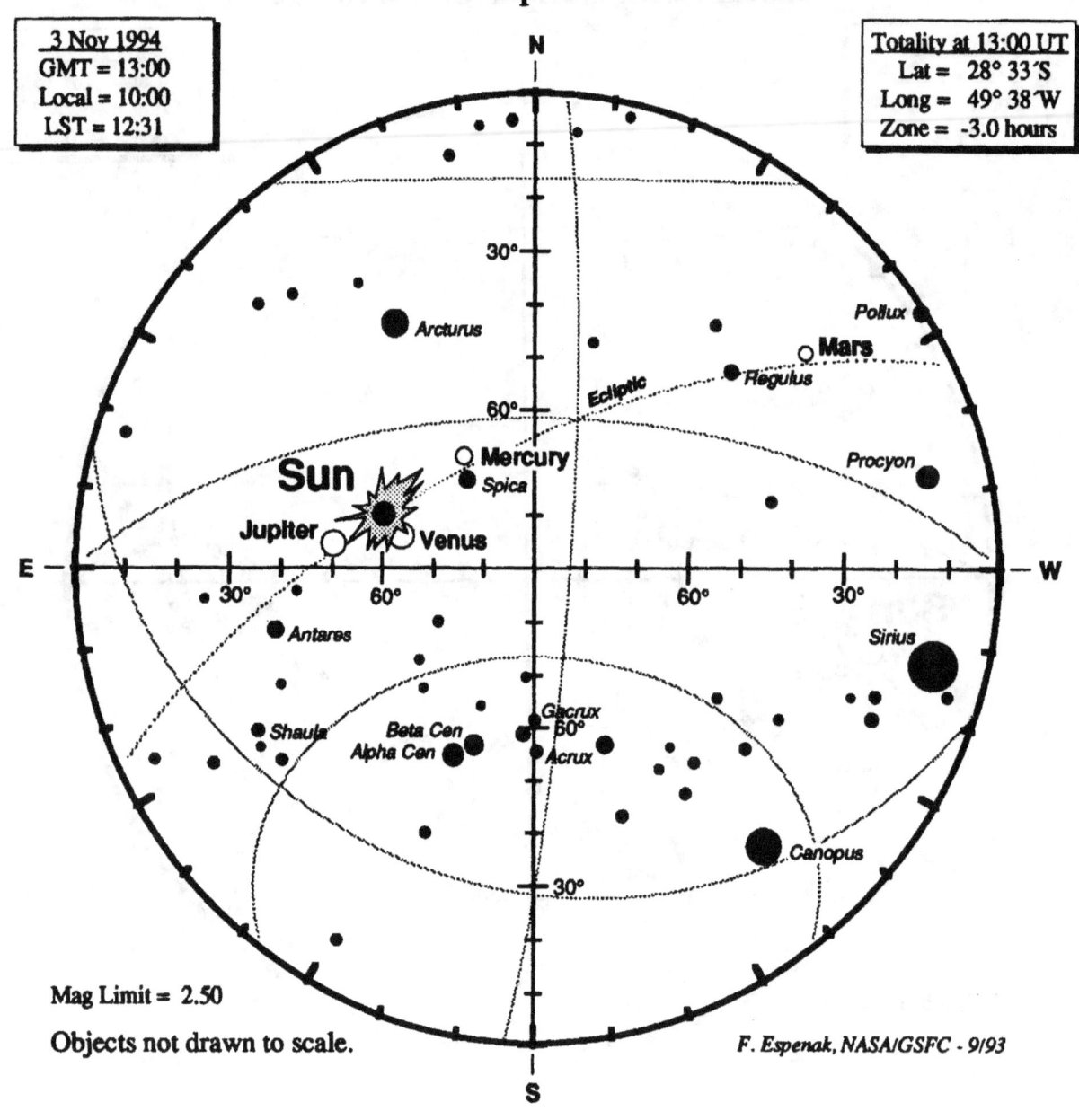

TOTAL SOLAR ECLIPSE OF 3 NOV 1994

TABLES

Table 1

ELEMENTS OF THE TOTAL SOLAR ECLIPSE OF 3 NOVEMBER 1994

Geocentric Conjunction   13:48:07.57 TDT   J.D. = 2449660.075088
of Sun & Moon in R.A.:   (=13:47:07.87 UT)

Instant of   13:40:06.02 TDT   J.D. = 2449660.069514
Greatest Eclipse: (=13:39:06.32 UT)

Geocentric Coordinates of Sun & Moon at Greatest Eclipse (DE200/LE200):

Sun:	R.A. = 14h 33m 55.825s		Moon:	R.A. = 14h 33m 36.503s
	Dec. = $-15°-05'-51.16"$			Dec. = $-15°-26'-53.71"$
	Semi-Diameter = 16' 07.43"			Semi-Diameter = 16' 43.07"
	Eq.Hor.Par. = 8.87"			Eq.Hor.Par. = 1° 01' 21.06"
	Δ R.A. = 9.863s/h			Δ R.A. = 154.292s/h
	Δ Dec. = $-46.79"$/h			Δ Dec. = $-509.87"$/h

Lunar Radius   k1 = 0.2725076 (Penumbra)   Shift in Δb = 0.0"
Constants: k2 = 0.2722810 (Umbra)   Lunar Position: Δl = 0.0"

Geocentric Libration: l = $-0.8°$ Brown Lun. No. = 1173
(Optical + Physical)   b = 0.5°   Saros Series = 133 (44/72)
                         c = 19.7°   Ephemeris = (DE200/LE200)

Eclipse Magnitude = 1.05351   Gamma = -0.35216 ΔT = 59.7 s

Polynomial Besselian Elements for: 3 Nov 1994   14:00:00.0 TDT $(=t_0)$

n	x	y	d	l_1	l_2	μ
0	0.1125604	-0.3855728	-15.1009045	0.5366173	-0.0094926	394.101532
1	0.5687835	-0.1257806	-0.0126857	-0.0000313	-0.0000312	15.001422
2	0.0000207	0.0001233	0.0000033	-0.0000130	-0.0000130	-0.000003
3	-0.0000097	0.0000020				

Tan f_1 = 0.0047133   Tan f_2 = 0.0046898

At time 't_1' (decimal hours), each besselian element is evaluated by:

$$x = x_0 + x_1*t + x_2*t^2 + x_3*t^3 \quad (\text{or } x = \sum [x_n*t^n]; n = 0 \text{ to } 3)$$

where: $t = t_1 - t_0$ (decimal hours) and $t_0 = 17.000$

Note that all times are expressed in Terrestrial Dynamical Time (TDT).

Saros Series 133: Member 44 of 72 eclipses in series.

Table 2

SHADOW CONTACTS AND CIRCUMSTANCES
TOTAL SOLAR ECLIPSE OF 3 NOVEMBER 1994

ΔT = 59.7 s
= 000°15.0′E

		Terrestrial Dynamical Time h m s	Latitude	Ephemeris Longitude†	True Longitude*
External/Internal Contacts of Penumbra:	P1	11:06:00.2	00°43.1′S	080°47.5′W	080°32.6′W
	P2	13:09:15.5	35°49.7′S	122°38.4′W	122°23.4′W
	P3	14:10:44.1	58°51.5′S	079°44.6′E	079°59.5′E
	P4	16:14:07.0	24°53.0′S	029°34.2′E	029°49.2′E
Extreme North/South Limits of Penumbral Path:	N1	11:52:15.3	22°12.4′N	085°50.9′W	085°36.0′W
	S1	12:55:47.6	48°12.1′S	125°35.9′W	125°20.9′W
	N2	15:28:01.2	01°59.7′S	034°26.6′E	034°41.6′E
	S2	14:24:08.4	69°11.7′S	095°08.0′E	095°22.9′E
External/Internal Contacts of Umbra:	U1	12:02:38.6	07°41.6′S	096°50.7′W	096°35.7′W
	U2	12:04:43.9	08°05.5′S	097°28.6′W	097°13.6′W
	U3	15:15:21.6	32°10.2′S	046°50.4′E	047°05.3′E
	U4	15:17:28.0	31°46.5′S	046°09.8′E	046°24.8′E
Extreme North/South Limits of Umbral Path:	N1	12:03:13.5	07°19.8′S	096°53.5′W	096°38.5′W
	S1	12:04:09.8	08°27.1′S	097°26.0′W	097°11.0′W
	N2	15:16:53.0	31°24.9′S	046°10.4′E	046°25.3′E
	S2	15:15:55.9	32°31.7′S	046°50.0′E	047°05.0′E
Extreme Limits of Center Line:	C1	12:03:41.2	07°53.4′S	097°09.6′W	096°54.6′W
	C2	15:16:24.8	31°58.3′S	046°30.0′E	046°45.0′E
Instant of Greatest Eclipse:	G0	13:40:06.0	35°21.4′S	034°28.4′W	034°13.4′W

Circumstances at
Greatest Eclipse: Sun's Altitude = 69.1°   Path Width = 189.0 km
                  Sun's Azimuth = 15.3°   Central Duration = 04m21.9s

---

† Ephemeris Longitude is the terrestrial dynamical longitude assuming a uniformly rotating Earth.

* True Longitude is calculated by correcting the Ephemeris Longitude for the non-uniform rotation of Earth.
(T.L. = E.L. - 1.002738*ΔT/240, where ΔT (in seconds) = TDT - UT)

Note: Longitude is measured positive to the East.

Since ΔT is not known in advance, the value used in the predictions is an extrapolation based on pre-1992 measurements. Nevertheless, the actual value is expected to fall within ±0.3 seconds of the estimated ΔT used here.

Table 3

PATH OF THE UMBRAL SHADOW
TOTAL SOLAR ECLIPSE OF 3 NOVEMBER 1994

Universal Time	Northern Limit Latitude Longitude	Southern Limit Latitude Longitude	Center Line Latitude Longitude	Path Azm °	Path Width km	Central Durat.
Limits	07°19.8'S 096°38.5'W	08°27.1'S 097°11.0'W	07°53.4'S 096°54.6'W	—	135	01m52s
12:05	11°02.9'S 084°36.8'W	11°25.1'S 087°40.3'W	11°15.6'S 086°04.0'W	107	151	02m14s
12:10	13°51.0'S 077°18.6'W	14°34.9'S 079°23.1'W	14°13.5'S 078°19.6'W	111	162	02m33s
12:15	15°56.8'S 072°28.0'W	16°49.0'S 074°16.1'W	16°23.3'S 073°21.3'W	114	169	02m47s
12:20	17°44.0'S 068°38.6'W	18°41.6'S 070°18.2'W	18°13.1'S 069°27.7'W	116	174	02m59s
12:25	19°19.6'S 065°24.0'W	20°21.4'S 066°58.1'W	19°50.8'S 066°10.4'W	118	178	03m10s
12:30	20°47.2'S 062°32.1'W	21°52.5'S 064°02.2'W	21°20.1'S 063°16.6'W	119	181	03m19s
12:35	22°08.5'S 059°56.2'W	23°16.9'S 061°23.0'W	22°42.9'S 060°39.1'W	120	183	03m28s
12:40	23°24.9'S 057°31.9'W	24°36.1'S 058°55.8'W	24°00.7'S 058°13.4'W	120	185	03m36s
12:45	24°37.0'S 055°16.3'W	25°50.8'S 056°37.5'W	25°14.1'S 055°56.5'W	121	187	03m43s
12:50	25°45.5'S 053°07.2'W	27°01.8'S 054°25.9'W	26°23.7'S 053°46.1'W	121	188	03m50s
12:55	26°50.7'S 051°03.0'W	28°09.3'S 052°19.1'W	27°30.1'S 051°40.7'W	121	189	03m56s
13:00	27°53.0'S 049°02.5'W	29°13.7'S 050°16.0'W	28°33.4'S 049°38.8'W	120	189	04m01s
13:05	28°52.5'S 047°04.5'W	30°15.4'S 048°15.3'W	29°34.0'S 047°39.5'W	120	190	04m06s
13:10	29°49.4'S 045°08.2'W	31°14.3'S 046°16.1'W	30°31.9'S 045°41.8'W	120	190	04m11s
13:15	30°43.9'S 043°12.8'W	32°10.8'S 044°17.7'W	31°27.4'S 043°44.9'W	119	190	04m14s
13:20	31°36.1'S 041°17.8'W	33°04.7'S 042°19.4'W	32°20.4'S 041°48.3'W	118	190	04m17s
13:25	32°25.9'S 039°22.3'W	33°56.3'S 040°20.5'W	33°11.1'S 039°51.1'W	117	190	04m20s
13:30	33°13.4'S 037°26.0'W	34°45.4'S 038°20.4'W	33°59.4'S 037°53.0'W	116	190	04m22s
13:35	33°58.6'S 035°28.3'W	35°32.2'S 036°18.7'W	34°45.4'S 035°53.3'W	115	189	04m23s
13:40	34°41.5'S 033°28.6'W	36°16.5'S 034°14.6'W	35°29.0'S 033°51.4'W	114	189	04m23s
13:45	35°22.1'S 031°26.4'W	36°58.4'S 032°07.8'W	36°10.2'S 031°47.0'W	112	188	04m23s
13:50	36°00.2'S 029°21.3'W	37°37.7'S 029°57.6'W	36°48.9'S 029°39.4'W	111	188	04m22s
13:55	36°35.9'S 027°12.6'W	38°14.3'S 027°43.6'W	37°25.0'S 027°28.1'W	109	187	04m21s
14:00	37°08.9'S 024°59.9'W	38°48.1'S 025°25.2'W	37°58.4'S 025°12.6'W	107	186	04m19s
14:05	37°39.1'S 022°42.6'W	39°18.9'S 023°01.6'W	38°28.9'S 022°52.2'W	105	186	04m16s
14:10	38°06.4'S 020°19.9'W	39°46.5'S 020°32.4'W	38°56.4'S 020°26.4'W	104	185	04m13s
14:15	38°30.6'S 017°51.3'W	40°10.7'S 017°56.7'W	39°20.6'S 017°54.3'W	102	184	04m08s
14:20	38°51.3'S 015°15.7'W	40°31.2'S 015°13.7'W	39°41.2'S 015°15.1'W	100	183	04m04s
14:25	39°08.3'S 012°32.4'W	40°47.7'S 012°22.4'W	39°57.9'S 012°27.8'W	97	181	03m58s
14:30	39°21.2'S 009°39.9'W	40°59.6'S 009°21.5'W	40°10.4'S 009°31.3'W	95	180	03m52s
14:35	39°29.4'S 006°37.0'W	41°06.5'S 006°09.6'W	40°17.9'S 006°24.0'W	93	178	03m45s
14:40	39°32.4'S 003°21.7'W	41°07.5'S 002°44.6'W	40°20.0'S 003°03.9'W	91	176	03m37s
14:45	39°29.3'S 000°08.8'E	41°01.9'S 000°56.2'E	40°15.6'S 000°31.6'E	88	174	03m29s
14:50	39°18.8'S 003°57.9'E	40°48.1'S 004°56.8'E	40°03.5'S 004°26.4'E	86	172	03m20s
14:55	38°59.3'S 008°11.4'E	40°24.2'S 009°23.2'E	39°41.9'S 008°46.2'E	84	169	03m09s
15:00	38°27.8'S 012°58.6'E	39°46.9'S 014°25.9'E	39°07.6'S 013°40.9'E	81	166	02m58s
15:05	37°39.1'S 018°37.2'E	38°49.9'S 020°25.2'E	38°14.9'S 019°29.5'E	79	161	02m45s
15:10	36°20.8'S 025°50.4'E	37°17.2'S 028°15.1'E	36°49.8'S 027°00.1'E	77	155	02m29s
15:15	33°32.3'S 038°16.2'E	00°00.0'N 000°00.0'E	33°28.0'S 041°11.3'E	74	142	02m02s
Limits	31°24.9'S 046°25.3'E	32°31.7'S 047°05.0'E	31°58.3'S 046°45.0'E	—	137	01m53s

Table 4

PHYSICAL EPHEMERIS OF THE UMBRAL SHADOW
TOTAL SOLAR ECLIPSE OF 3 NOVEMBER 1994

Universal Time	Center Line Latitude	Center Line Longitude	Diameter Ratio	Eclipse Obscur.	Sun Alt °	Sun Azm °	Path Width km	Major Axis km	Minor Axis km	Umbra Veloc. km/s	Central Durat.
12:02.7	07°53.4'S	096°54.6'W	1.0360	1.0732	0.0	105.2	135.3	–	121.0	–	01m52s
12:05	11°15.6'S	086°04.0'W	1.0397	1.0810	11.7	103.3	150.7	654.9	133.1	4.320	02m14s
12:10	14°13.5'S	078°19.6'W	1.0426	1.0870	20.9	101.0	162.0	398.5	142.3	2.285	02m33s
12:15	16°23.3'S	073°21.3'W	1.0445	1.0909	27.2	98.8	169.0	324.3	148.2	1.701	02m47s
12:20	18°13.1'S	069°27.7'W	1.0459	1.0939	32.3	96.7	174.0	286.0	152.9	1.402	02m59s
12:25	19°50.8'S	066°10.4'W	1.0471	1.0964	36.8	94.3	177.9	261.9	156.7	1.216	03m10s
12:30	21°20.1'S	063°16.6'W	1.0481	1.0985	40.7	91.9	180.9	245.2	159.9	1.088	03m19s
12:35	22°42.9'S	060°39.1'W	1.0490	1.1004	44.3	89.2	183.3	232.9	162.6	0.994	03m28s
12:40	24°00.7'S	058°13.4'W	1.0497	1.1020	47.6	86.2	185.1	223.4	165.0	0.923	03m36s
12:45	25°14.1'S	055°56.5'W	1.0504	1.1034	50.7	83.0	186.6	216.0	167.1	0.867	03m43s
12:50	26°23.7'S	053°46.1'W	1.0510	1.1046	53.6	79.4	187.7	210.0	169.0	0.823	03m50s
12:55	27°30.1'S	051°40.7'W	1.0515	1.1057	56.3	75.5	188.6	205.2	170.6	0.787	03m56s
13:00	28°33.4'S	049°38.8'W	1.0519	1.1066	58.7	71.0	189.2	201.2	172.0	0.758	04m01s
13:05	29°34.0'S	047°39.5'W	1.0523	1.1074	61.0	66.0	189.6	198.0	173.2	0.734	04m06s
13:10	30°31.9'S	045°41.8'W	1.0526	1.1081	63.1	60.4	189.9	195.4	174.2	0.716	04m11s
13:15	31°27.4'S	043°44.9'W	1.0529	1.1086	64.9	54.2	190.0	193.4	175.0	0.701	04m14s
13:20	32°20.4'S	041°48.3'W	1.0531	1.1091	66.4	47.2	190.0	191.8	175.7	0.689	04m17s
13:25	33°11.1'S	039°51.1'W	1.0533	1.1094	67.6	39.5	189.9	190.6	176.2	0.681	04m20s
13:30	33°59.4'S	037°53.0'W	1.0534	1.1097	68.5	31.1	189.6	189.8	176.6	0.676	04m22s
13:35	34°45.4'S	035°53.3'W	1.0535	1.1098	69.1	22.2	189.3	189.3	176.8	0.673	04m23s
13:40	35°29.0'S	033°51.4'W	1.0535	1.1099	69.2	13.0	188.9	189.3	176.9	0.673	04m23s
13:45	36°10.2'S	031°47.0'W	1.0535	1.1098	68.9	3.8	188.4	189.5	176.8	0.676	04m23s
13:50	36°48.9'S	029°39.4'W	1.0534	1.1097	68.2	354.9	187.9	190.2	176.6	0.681	04m22s
13:55	37°25.0'S	027°28.1'W	1.0533	1.1094	67.2	346.5	187.2	191.2	176.2	0.689	04m21s
14:00	37°58.4'S	025°12.6'W	1.0531	1.1091	65.8	338.6	186.5	192.6	175.7	0.701	04m19s
14:05	38°28.9'S	022°52.2'W	1.0529	1.1086	64.2	331.4	185.6	194.4	175.0	0.716	04m16s
14:10	38°56.4'S	020°26.4'W	1.0526	1.1080	62.3	324.7	184.7	196.7	174.1	0.734	04m13s
14:15	39°20.6'S	017°54.3'W	1.0523	1.1073	60.1	318.6	183.7	199.6	173.1	0.757	04m08s
14:20	39°41.2'S	015°15.1'W	1.0519	1.1065	57.8	312.9	182.5	203.2	171.9	0.785	04m04s
14:25	39°57.9'S	012°27.8'W	1.0515	1.1055	55.2	307.6	181.2	207.5	170.4	0.820	03m58s
14:30	40°10.4'S	009°31.3'W	1.0509	1.1045	52.5	302.6	179.8	212.8	168.8	0.862	03m52s
14:35	40°17.9'S	006°24.0'W	1.0503	1.1032	49.5	297.9	178.2	219.4	166.9	0.913	03m45s
14:40	40°20.0'S	003°03.9'W	1.0496	1.1018	46.4	293.3	176.4	227.6	164.7	0.978	03m37s
14:45	40°15.6'S	000°31.6'E	1.0489	1.1001	43.0	288.8	174.3	238.2	162.2	1.061	03m29s
14:50	40°03.5'S	004°26.4'E	1.0479	1.0982	39.2	284.3	171.9	252.1	159.4	1.171	03m20s
14:55	39°41.9'S	008°46.2'E	1.0469	1.0960	35.1	279.8	169.1	271.4	156.0	1.322	03m09s
15:00	39°07.6'S	013°40.9'E	1.0456	1.0933	30.5	275.1	165.6	300.0	152.0	1.547	02m58s
15:05	38°14.9'S	019°29.5'E	1.0441	1.0901	25.0	270.1	161.2	348.2	147.0	1.927	02m45s
15:10	36°49.8'S	027°00.1'E	1.0420	1.0857	18.0	264.3	155.0	454.8	140.2	2.767	02m29s
15:15	33°28.0'S	041°11.3'E	1.0378	1.0771	5.0	255.2	141.9	1473.6	127.0	10.821	02m02s
15:15.4	31°58.3'S	046°45.0'E	1.0363	1.0738	0.0	252.1	136.5	–	121.9	–	01m53s

Table 5

LOCAL CIRCUMSTANCES ON THE CENTER LINE
TOTAL SOLAR ECLIPSE OF 3 NOVEMBER 1994

Center Line Maximum Eclipse			First Contact				Second Contact			Third Contact			Fourth Contact			
U.T.	Durat.	Alt °	U.T.	P °	V °	Alt °	U.T.	P °	V °	U.T.	P °	V °	U.T.	P °	V °	Alt °
12:05	02m14s	12	-	-	-	-	12:03:53	110	209	12:06:07	290	29	13:05:33	111	206	26
12:10	02m33s	21	11:11:58	290	33	7	12:08:44	111	211	12:11:17	291	31	13:15:04	112	209	37
12:15	02m47s	27	11:14:25	291	35	13	12:13:37	112	213	12:16:24	292	33	13:23:18	113	211	43
12:20	02m59s	32	11:17:14	291	36	18	12:18:31	112	214	12:21:30	292	34	13:31:00	113	213	49
12:25	03m10s	37	11:20:17	292	37	22	12:23:25	113	216	12:26:35	293	36	13:38:20	113	216	54
12:30	03m19s	41	11:23:30	292	39	25	12:28:21	113	218	12:31:40	293	38	13:45:23	113	219	58
12:35	03m28s	44	11:26:50	292	40	29	12:33:16	113	220	12:36:44	293	40	13:52:12	113	223	62
12:40	03m36s	48	11:30:16	292	42	32	12:38:12	113	222	12:41:48	293	42	13:58:48	113	228	65
12:45	03m43s	51	11:33:48	293	43	35	12:43:09	113	224	12:46:52	293	45	14:05:13	113	234	68
12:50	03m50s	54	11:37:25	293	44	37	12:48:05	113	227	12:51:55	293	47	14:11:26	113	241	71
12:55	03m56s	56	11:41:06	293	46	40	12:53:02	113	230	12:56:58	293	50	14:17:29	113	250	73
13:00	04m01s	59	11:44:53	293	48	43	12:58:00	113	233	13:02:01	293	54	14:23:21	112	260	74
13:05	04m06s	61	11:48:44	293	49	45	13:02:57	113	237	13:07:03	293	57	14:29:03	112	271	74
13:10	04m11s	63	11:52:39	293	51	47	13:07:55	112	241	13:12:06	292	62	14:34:36	112	282	74
13:15	04m14s	65	11:56:40	292	53	50	13:12:53	112	246	13:17:07	292	67	14:39:59	111	292	74
13:20	04m17s	66	12:00:45	292	56	52	13:17:52	112	251	13:22:09	292	72	14:45:12	111	301	72
13:25	04m20s	68	12:04:56	292	58	54	13:22:50	112	257	13:27:10	291	79	14:50:17	110	308	71
13:30	04m22s	69	12:09:12	292	61	56	13:27:49	111	264	13:32:11	291	86	14:55:12	110	314	69
13:35	04m23s	69	12:13:34	291	64	58	13:32:49	111	271	13:37:11	291	93	14:59:59	109	318	67
13:40	04m23s	69	12:18:02	291	68	60	13:37:48	110	278	13:42:12	290	100	15:04:38	109	322	65
13:45	04m23s	69	12:22:36	291	72	61	13:42:48	110	285	13:47:12	290	108	15:09:09	108	324	63
13:50	04m22s	68	12:27:18	290	76	63	13:47:49	109	292	13:52:11	289	114	15:13:33	107	326	60
13:55	04m21s	67	12:32:06	290	81	64	13:52:50	109	299	13:57:10	288	121	15:17:49	107	328	58
14:00	04m19s	66	12:37:02	289	87	65	13:57:51	108	304	14:02:09	288	126	15:21:58	106	329	55
14:05	04m16s	64	12:42:06	289	93	66	14:02:52	107	309	14:07:08	287	131	15:26:00	105	330	53
14:10	04m13s	62	12:47:20	288	100	66	14:07:54	107	314	14:12:06	287	135	15:29:56	105	331	50
14:15	04m08s	60	12:52:42	288	106	66	14:12:56	106	318	14:17:04	286	139	15:33:45	104	332	48
14:20	04m04s	58	12:58:14	287	113	65	14:17:58	105	321	14:22:02	285	142	15:37:29	104	332	45
14:25	03m58s	55	13:03:57	286	120	64	14:23:01	105	323	14:26:59	284	144	15:41:06	103	332	42
14:30	03m52s	52	13:09:50	285	126	63	14:28:04	104	326	14:31:56	284	146	15:44:37	102	333	39
14:35	03m45s	50	13:15:56	285	132	61	14:33:07	103	327	14:36:52	283	148	15:48:02	102	333	36
14:40	03m37s	46	13:22:14	284	137	59	14:38:11	102	329	14:41:48	282	149	15:51:21	101	333	33
14:45	03m29s	43	13:28:46	283	141	56	14:43:15	102	330	14:46:44	282	150	15:54:33	100	333	30
14:50	03m20s	39	13:35:33	282	145	52	14:48:20	101	331	14:51:40	281	151	15:57:37	100	332	26
14:55	03m09s	35	13:42:38	281	148	48	14:53:25	100	332	14:56:34	280	152	16:00:32	99	332	23
15:00	02m58s	30	13:50:02	280	150	44	14:58:31	99	333	15:01:29	279	153	16:03:16	99	332	18
15:05	02m45s	25	13:57:55	280	153	38	15:03:37	99	333	15:06:22	279	153	16:05:42	98	332	13
15:10	02m29s	18	14:06:32	278	154	31	15:08:45	98	334	15:11:14	278	154	16:07:38	97	331	7
15:15	02m02s	5	14:17:46	277	156	17	15:13:59	97	334	15:16:01	277	153	-	-	-	-

Table 6

TOPOCENTRIC DATA AND PATH CORRECTIONS DUE TO LUNAR LIMB PROFILE

Universal Time	Moon Topo H.P. "	Moon Topo S.D. "	Moon Rel. Ang.V "/s	Topo Lib. Long °	Sun Alt. °	Sun Az. °	Path Az. °	North Limit P.A. °	North Limit Int. '	North Limit Ext. '	South Limit Int. '	South Limit Ext. '	Central Durat. Cor. s
12:05	3694.0	1005.9	0.574	-0.01	11.7	103.3	109.7	20.0	-0.1	0.4	0.7	-2.2	-2.3
12:10	3704.2	1008.6	0.538	-0.05	20.9	101.0	113.2	21.1	-0.1	0.9	0.6	-2.8	-2.6
12:15	3710.9	1010.4	0.514	-0.09	27.2	98.8	115.4	21.8	-0.1	1.0	0.5	-3.1	-2.8
12:20	3716.1	1011.8	0.496	-0.13	32.3	96.7	117.0	22.3	0.0	1.1	0.5	-3.3	-2.9
12:25	3720.3	1013.0	0.480	-0.17	36.8	94.3	118.3	22.6	0.0	1.1	0.4	-3.3	-3.1
12:30	3724.0	1014.0	0.467	-0.22	40.7	91.9	119.2	22.9	0.0	1.1	0.4	-3.3	-3.2
12:35	3727.1	1014.8	0.456	-0.26	44.3	89.2	119.9	23.0	0.0	1.1	0.4	-3.3	-3.3
12:40	3729.9	1015.6	0.446	-0.30	47.6	86.2	120.4	23.1	0.1	1.1	0.3	-3.3	-3.5
12:45	3732.3	1016.2	0.437	-0.34	50.7	83.0	120.6	23.2	0.1	1.1	0.3	-3.3	-3.6
12:50	3734.4	1016.8	0.429	-0.39	53.6	79.4	120.7	23.1	0.1	1.1	0.3	-3.3	-3.7
12:55	3736.2	1017.3	0.422	-0.43	56.3	75.5	120.6	23.1	0.1	1.1	0.3	-3.3	-3.7
13:00	3737.8	1017.7	0.416	-0.47	58.7	71.0	120.3	22.9	0.0	1.1	0.4	-3.3	-3.8
13:05	3739.2	1018.1	0.411	-0.51	61.0	66.0	119.9	22.7	0.0	1.0	0.3	-3.3	-3.8
13:10	3740.4	1018.4	0.407	-0.56	63.1	60.4	119.3	22.5	0.0	1.0	0.4	-3.2	-3.9
13:15	3741.4	1018.7	0.403	-0.60	64.9	54.2	118.6	22.2	0.0	1.0	0.4	-3.2	-3.9
13:20	3742.2	1018.9	0.400	-0.64	66.4	47.2	117.7	21.9	0.0	1.0	0.4	-3.1	-3.8
13:25	3742.8	1019.0	0.397	-0.69	67.6	39.5	116.7	21.5	-0.1	0.9	0.5	-2.9	-3.8
13:30	3743.2	1019.1	0.395	-0.73	68.5	31.1	115.6	21.1	-0.1	0.8	0.4	-2.7	-3.5
13:35	3743.4	1019.2	0.394	-0.77	69.1	22.2	114.3	20.6	-0.1	0.6	0.4	-2.4	-3.4
13:40	3743.5	1019.2	0.393	-0.82	69.2	13.0	112.9	20.2	-0.1	0.3	0.5	-2.1	-3.3
13:45	3743.4	1019.2	0.393	-0.86	68.9	3.8	111.5	19.6	-0.2	0.3	0.5	-2.2	-3.3
13:50	3743.2	1019.1	0.394	-0.90	68.2	354.9	109.9	19.1	-0.2	0.4	0.5	-2.2	-3.1
13:55	3742.7	1019.0	0.395	-0.95	67.2	346.5	108.2	18.5	-0.2	0.3	0.6	-2.0	-3.2
14:00	3742.1	1018.9	0.397	-0.99	65.8	338.6	106.4	17.9	-0.2	0.1	0.6	-1.8	-3.3
14:05	3741.3	1018.6	0.400	-1.03	64.2	331.4	104.5	17.3	-0.3	0.0	0.7	-1.4	-3.4
14:10	3740.3	1018.4	0.403	-1.08	62.3	324.7	102.6	16.6	-0.3	0.0	0.8	-1.5	-3.4
14:15	3739.1	1018.1	0.407	-1.12	60.1	318.6	100.6	15.9	-0.3	0.3	0.8	-1.6	-3.3
14:20	3737.7	1017.7	0.412	-1.16	57.8	312.9	98.5	15.2	-0.3	0.4	0.9	-2.0	-3.3
14:25	3736.1	1017.2	0.418	-1.21	55.2	307.6	96.4	14.5	-0.3	0.5	0.9	-2.2	-3.2
14:30	3734.2	1016.7	0.425	-1.25	52.5	302.7	94.2	13.8	-0.2	0.5	0.9	-2.3	-3.1
14:35	3732.1	1016.2	0.433	-1.29	49.5	297.9	91.9	13.1	-0.3	0.6	1.0	-2.2	-2.9
14:40	3729.6	1015.5	0.442	-1.33	46.4	293.3	89.6	12.4	-0.4	0.5	1.0	-1.9	-2.7
14:45	3726.8	1014.7	0.453	-1.38	43.0	288.8	87.3	11.6	-0.5	0.4	1.0	-1.5	-2.5
14:50	3723.5	1013.9	0.465	-1.42	39.2	284.3	85.0	10.9	-0.5	0.5	1.1	-1.0	-2.3
14:55	3719.7	1012.8	0.479	-1.46	35.1	279.8	82.6	10.1	-0.6	0.6	1.1	-0.9	-2.1
15:00	3715.2	1011.6	0.496	-1.50	30.5	275.1	80.3	9.4	-0.8	0.7	1.1	-1.7	-1.8
15:05	3709.6	1010.1	0.517	-1.54	25.0	270.1	77.9	8.6	-0.8	0.8	1.1	-2.3	-1.6
15:10	3702.0	1008.0	0.546	-1.58	18.0	264.3	75.4	7.8	-0.9	0.6	1.1	-2.7	-1.3
15:15	3687.5	1004.1	0.601	-1.62	5.0	255.2	0.0	6.7	0.0	0.0	0.0	0.0	-1.1

Table 7
MAPPING COORDINATES FOR THE UMBRAL PATH

Longitude	Latitude of: Northern Limit	Latitude of: Southern Limit	Latitude of: Center Line	Universal Time at: Northern Limit	Universal Time at: Southern Limit	Universal Time at: Center Line	Circumstances on the Center Line Sun Alt °	Circumstances on the Center Line Sun Az. °	Circumstances on the Center Line Path Width km	Circumstances on the Center Line Center Durat.
097°00.0′W	–	08°30.1′S	–	–	12:02:37	–				
096°00.0′W	07°14.8′S	08°48.0′S	08°08.4′S	12:02:10	12:03:16	12:02:42	1	105	137	01m54s
095°00.0′W	07°48.6′S	09°05.0′S	08°25.3′S	12:02:17	12:03:16	12:02:45	2	105	138	01m55s
094°00.0′W	08°03.4′S	09°22.1′S	08°42.6′S	12:02:20	12:03:22	12:02:51	3	105	139	01m57s
093°00.0′W	08°20.6′S	09°40.4′S	09°00.3′S	12:02:27	12:03:31	12:02:58	4	105	141	01m59s
092°00.0′W	08°38.3′S	09°59.1′S	09°18.5′S	12:02:36	12:03:42	12:03:08	5	104	142	02m01s
091°00.0′W	08°56.5′S	10°18.2′S	09°37.1′S	12:02:47	12:03:56	12:03:21	6	104	144	02m03s
090°00.0′W	09°15.0′S	10°37.8′S	09°56.2′S	12:03:01	12:04:12	12:03:36	7	104	145	02m05s
089°00.0′W	09°34.0′S	10°57.8′S	10°15.7′S	12:03:17	12:04:31	12:03:53	8	104	146	02m07s
088°00.0′W	09°53.5′S	11°18.3′S	10°35.6′S	12:03:36	12:04:52	12:04:13	10	104	148	02m10s
087°00.0′W	10°13.4′S	11°39.3′S	10°56.1′S	12:03:57	12:05:17	12:04:36	11	103	149	02m12s
086°00.0′W	10°33.8′S	12°00.7′S	11°17.0′S	12:04:22	12:05:44	12:05:02	12	103	151	02m14s
085°00.0′W	10°54.6′S	12°22.6′S	11°38.4′S	12:04:49	12:06:14	12:05:30	13	103	152	02m16s
084°00.0′W	11°16.0′S	12°45.1′S	12°00.2′S	12:05:19	12:06:47	12:06:02	14	103	154	02m19s
083°00.0′W	11°37.8′S	13°08.0′S	12°22.6′S	12:05:52	12:07:23	12:06:36	15	102	155	02m21s
082°00.0′W	12°00.1′S	13°31.4′S	12°45.4′S	12:06:28	12:08:02	12:07:14	16	102	157	02m24s
081°00.0′W	12°22.8′S	13°55.3′S	13°08.8′S	12:07:07	12:08:44	12:07:55	18	102	158	02m26s
080°00.0′W	12°46.1′S	14°19.7′S	13°32.6′S	12:07:49	12:09:30	12:08:39	19	102	160	02m29s
079°00.0′W	13°09.8′S	14°44.6′S	13°56.9′S	12:08:35	12:10:19	12:09:26	20	101	161	02m31s
078°00.0′W	13°34.0′S	15°10.0′S	14°21.7′S	12:09:24	12:11:12	12:10:17	21	101	162	02m34s
077°00.0′W	13°58.7′S	15°35.8′S	14°47.0′S	12:10:17	12:12:08	12:11:11	23	100	164	02m37s
076°00.0′W	14°24.0′S	16°02.2′S	15°12.8′S	12:11:13	12:13:08	12:12:09	24	100	165	02m40s
075°00.0′W	14°49.6′S	16°29.1′S	15°39.0′S	12:12:12	12:14:11	12:13:11	25	100	167	02m42s
074°00.0′W	15°15.8′S	16°56.4′S	16°05.8′S	12:13:16	12:15:18	12:14:16	26	99	168	02m45s
073°00.0′W	15°42.5′S	17°24.2′S	16°33.0′S	12:14:23	12:16:29	12:15:25	28	99	169	02m48s
072°00.0′W	16°09.6′S	17°52.5′S	17°00.7′S	12:15:34	12:17:44	12:16:38	29	98	171	02m51s
071°00.0′W	16°37.2′S	18°21.3′S	17°28.9′S	12:16:48	12:19:03	12:17:54	30	98	172	02m54s
070°00.0′W	17°05.2′S	18°50.4′S	17°57.5′S	12:18:07	12:20:25	12:19:15	32	97	173	02m58s
069°00.0′W	17°33.7′S	19°20.0′S	18°26.5′S	12:19:30	12:21:52	12:20:40	33	96	175	03m01s
068°00.0′W	18°02.6′S	19°50.0′S	18°56.0′S	12:20:56	12:23:22	12:22:08	34	96	176	03m04s
067°00.0′W	18°31.9′S	20°20.4′S	19°25.8′S	12:22:27	12:24:57	12:23:41	36	95	177	03m07s
066°00.0′W	19°01.6′S	20°51.2′S	19°56.1′S	12:24:01	12:26:35	12:25:17	37	94	178	03m10s
065°00.0′W	19°31.7′S	21°22.3′S	20°26.6′S	12:25:40	12:28:18	12:26:58	38	93	179	03m14s
064°00.0′W	20°02.1′S	21°53.6′S	20°57.5′S	12:27:22	12:30:04	12:28:42	40	93	180	03m17s
063°00.0′W	20°32.8′S	22°25.3′S	21°28.7′S	12:29:09	12:31:54	12:30:30	41	92	181	03m20s
062°00.0′W	21°03.8′S	22°57.2′S	22°00.2′S	12:31:60	12:33:48	12:32:23	42	91	182	03m23s
061°00.0′W	21°35.1′S	23°29.3′S	22°31.9′S	12:32:54	12:35:45	12:34:19	44	90	183	03m27s
060°00.0′W	22°06.5′S	24°01.5′S	23°03.7′S	12:34:52	12:37:47	12:36:18	45	88	184	03m30s
059°00.0′W	22°38.2′S	24°33.9′S	23°35.7′S	12:36:54	12:39:51	12:38:22	47	87	185	03m33s
058°00.0′W	23°10.0′S	25°06.3′S	24°07.8′S	12:39:00	12:41:59	12:40:29	48	86	185	03m37s
057°00.0′W	23°41.8′S	25°38.7′S	24°40.0′S	12:41:09	12:44:10	12:42:39	49	85	186	03m40s
056°00.0′W	24°13.8′S	26°11.1′S	25°12.2′S	12:43:22	12:46:24	12:44:52	51	83	187	03m43s
055°00.0′W	24°45.7′S	26°43.4′S	25°44.3′S	12:45:37	12:48:41	12:47:08	52	82	187	03m46s
054°00.0′W	25°17.5′S	27°15.6′S	26°16.4′S	12:47:56	12:51:00	12:49:27	53	80	188	03m49s
053°00.0′W	25°49.3′S	27°47.6′S	26°48.2′S	12:50:17	12:53:22	12:51:49	55	78	188	03m52s
052°00.0′W	26°20.9′S	28°19.4′S	27°19.9′S	12:52:41	12:55:46	12:54:13	56	76	188	03m55s
051°00.0′W	26°52.3′S	28°50.9′S	27°51.4′S	12:55:07	12:58:12	12:56:39	57	74	189	03m58s
050°00.0′W	27°23.4′S	29°22.0′S	28°22.5′S	12:57:36	13:00:39	12:59:07	58	72	189	04m00s

Table 7
MAPPING COORDINATES FOR THE UMBRAL PATH

Longitude	Latitude of: Northern Limit	Latitude of: Southern Limit	Latitude of: Center Line	Universal Time at: Northern Limit h m s	Universal Time at: Southern Limit h m s	Universal Time at: Center Line h m s	Circumstances on the Center Line Sun Alt °	Circumstances on the Center Line Sun Az. °	Circumstances on the Center Line Path Width km	Circumstances on the Center Line Center Durat.
049°00.0´W	27°54.2´S	29°52.7´S	28°53.3´S	13:00:06	13:03:08	13:01:37	59	69	189	04m03s
048°00.0´W	28°24.7´S	30°23.0´S	29°23.7´S	13:02:38	13:05:38	13:04:08	61	67	190	04m05s
047°00.0´W	28°54.7´S	30°52.8´S	29°53.7´S	13:05:11	13:08:09	13:06:40	62	64	190	04m08s
046°00.0´W	29°24.3´S	31°22.1´S	30°23.1´S	13:07:46	13:10:41	13:09:13	63	61	190	04m10s
045°00.0´W	29°53.4´S	31°50.9´S	30°52.0´S	13:10:21	13:13:13	13:11:47	64	58	190	04m12s
044°00.0´W	30°21.9´S	32°19.0´S	31°20.4´S	13:12:57	13:15:45	13:14:21	65	55	190	04m14s
043°00.0´W	30°49.9´S	32°46.5´S	31°48.1´S	13:15:33	13:18:17	13:16:56	65	52	190	04m15s
042°00.0´W	31°17.2´S	33°13.3´S	32°15.2´S	13:18:10	13:20:49	13:19:30	66	48	190	04m17s
041°00.0´W	31°43.9´S	33°39.5´S	32°41.6´S	13:20:46	13:23:21	13:22:04	67	44	190	04m18s
040°00.0´W	32°09.9´S	34°04.9´S	33°07.3´S	13:23:22	13:25:51	13:24:37	68	40	190	04m20s
039°00.0´W	32°35.2´S	34°29.6´S	33°32.3´S	13:25:58	13:28:22	13:27:10	68	36	190	04m21s
038°00.0´W	32°59.8´S	34°53.5´S	33°56.6´S	13:28:33	13:30:51	13:29:42	68	32	190	04m21s
037°00.0´W	33°23.6´S	35°16.7´S	34°20.1´S	13:31:07	13:33:19	13:32:13	69	27	190	04m22s
036°00.0´W	33°46.7´S	35°39.1´S	34°42.9´S	13:33:40	13:35:46	13:34:43	69	23	189	04m23s
035°00.0´W	34°09.0´S	36°00.7´S	35°04.9´S	13:36:11	13:38:11	13:37:12	69	18	189	04m23s
034°00.0´W	34°30.6´S	36°21.5´S	35°26.0´S	13:38:42	13:40:35	13:39:39	69	14	189	04m23s
033°00.0´W	34°51.3´S	36°41.6´S	35°46.4´S	13:41:11	13:42:57	13:42:05	69	9	189	04m23s
032°00.0´W	35°11.3´S	37°00.8´S	36°06.0´S	13:43:38	13:45:18	13:44:29	69	5	188	04m23s
031°00.0´W	35°30.4´S	37°19.3´S	36°24.9´S	13:46:04	13:47:37	13:46:51	69	0	188	04m23s
030°00.0´W	35°48.8´S	37°37.0´S	36°42.9´S	13:48:28	13:49:55	13:49:12	68	356	188	04m23s
029°00.0´W	36°06.4´S	37°53.9´S	37°00.1´S	13:50:50	13:52:10	13:51:31	68	352	188	04m22s
028°00.0´W	36°23.2´S	38°10.0´S	37°16.6´S	13:53:11	13:54:24	13:53:48	67	348	187	04m21s
027°00.0´W	36°39.2´S	38°25.3´S	37°32.2´S	13:55:29	13:56:36	13:56:03	67	345	187	04m21s
026°00.0´W	36°54.4´S	38°39.9´S	37°47.2´S	13:57:45	13:58:45	13:58:16	66	341	187	04m20s
025°00.0´W	37°08.9´S	38°53.8´S	38°01.3´S	14:00:60	14:00:53	14:00:27	66	338	186	04m19s
024°00.0´W	37°22.6´S	39°06.8´S	38°14.7´S	14:02:12	14:02:59	14:02:37	65	335	186	04m17s
023°00.0´W	37°35.5´S	39°19.2´S	38°27.3´S	14:04:23	14:05:03	14:04:44	64	332	186	04m16s
022°00.0´W	37°47.7´S	39°30.8´S	38°39.2´S	14:06:31	14:07:05	14:06:49	63	329	185	04m15s
021°00.0´W	37°59.2´S	39°41.7´S	38°50.4´S	14:08:37	14:09:05	14:08:52	63	326	185	04m13s
020°00.0´W	38°09.9´S	39°51.9´S	39°00.9´S	14:10:41	14:11:04	14:10:53	62	324	185	04m12s
019°00.0´W	38°20.0´S	40°01.4´S	39°10.7´S	14:12:43	14:13:60	14:12:52	61	321	184	04m10s
018°00.0´W	38°29.3´S	40°10.3´S	39°19.7´S	14:14:43	14:14:54	14:14:49	60	319	184	04m09s
017°00.0´W	38°37.9´S	40°18.4´S	39°28.1´S	14:16:40	14:16:46	14:16:44	59	317	183	04m07s
016°00.0´W	38°45.9´S	40°25.9´S	39°35.9´S	14:18:36	14:18:36	14:18:37	58	314	183	04m05s
015°00.0´W	38°53.2´S	40°32.7´S	39°42.9´S	14:20:30	14:20:25	14:20:28	58	312	182	04m03s
014°00.0´W	38°59.8´S	40°38.9´S	39°49.3´S	14:22:21	14:22:11	14:22:17	57	310	182	04m01s
013°00.0´W	39°05.8´S	40°44.5´S	39°55.1´S	14:24:10	14:23:56	14:24:03	56	309	181	03m59s
012°00.0´W	39°11.1´S	40°49.5´S	40°00.2´S	14:25:58	14:25:38	14:25:48	55	307	181	03m57s
011°00.0´W	39°15.8´S	40°53.8´S	40°04.8´S	14:27:43	14:27:19	14:27:31	54	305	181	03m55s
010°00.0´W	39°20.0´S	40°57.5´S	40°08.7´S	14:29:26	14:28:58	14:29:12	53	303	180	03m53s
009°00.0´W	39°23.5´S	41°00.7´S	40°12.0´S	14:31:07	14:30:35	14:30:51	52	302	180	03m51s
008°00.0´W	39°26.4´S	41°03.2´S	40°14.7´S	14:32:46	14:32:10	14:32:29	51	300	179	03m49s
007°00.0´W	39°28.7´S	41°05.2´S	40°16.9´S	14:34:23	14:33:43	14:34:04	50	299	179	03m46s
006°00.0´W	39°30.5´S	41°06.7´S	40°18.5´S	14:35:58	14:35:15	14:35:37	49	297	178	03m44s
005°00.0´W	39°31.7´S	41°07.5´S	40°19.5´S	14:37:32	14:36:44	14:37:08	48	296	177	03m42s
004°00.0´W	39°32.3´S	41°07.9´S	40°20.0´S	14:39:03	14:38:12	14:38:38	47	295	177	03m39s
003°00.0´W	39°32.4´S	41°07.7´S	40°20.0´S	14:40:32	14:39:38	14:40:06	46	293	176	03m37s
002°00.0´W	39°32.0´S	41°06.9´S	40°19.4´S	14:41:59	14:41:03	14:41:31	45	292	176	03m35s
001°00.0´W	39°31.0´S	41°05.7´S	40°18.3´S	14:43:25	14:42:25	14:42:55	44	291	175	03m32s
000°00.0´E	39°29.5´S	41°04.0´S	40°16.7´S	14:44:48	14:43:46	14:44:17	43	289	175	03m30s

Table 7
MAPPING COORDINATES FOR THE UMBRAL PATH

Longitude	Latitude of:			Universal Time at:			Circumstances on the Center Line			
	Northern Limit	Southern Limit	Center Line	Northern Limit	Southern Limit	Center Line	Sun Alt °	Sun Az. °	Path Width km	Center Durat.
001° 00.0´E	39°27.6´S	41°01.7´S	40°14.5´S	14:46:09	14:45:05	14:45:38	42	288	174	03m28s
002° 00.0´E	39°25.1´S	40°59.0´S	40°11.9´S	14:47:29	14:46:22	14:46:56	42	287	173	03m25s
003° 00.0´E	39°22.1´S	40°55.7´S	40°08.8´S	14:48:47	14:47:38	14:48:13	41	286	173	03m23s
004° 00.0´E	39°18.7´S	40°52.0´S	40°05.3´S	14:50:03	14:48:52	14:49:28	40	285	172	03m21s
005° 00.0´E	39°14.8´S	40°47.9´S	40°01.2´S	14:51:17	14:50:04	14:50:41	39	284	172	03m18s
006° 00.0´E	39°10.4´S	40°43.2´S	39°56.7´S	14:52:29	14:51:14	14:51:52	38	283	171	03m16s
007° 00.0´E	39°05.6´S	40°38.1´S	39°51.8´S	14:53:39	14:52:23	14:53:01	37	282	170	03m14s
008° 00.0´E	39°00.3´S	40°32.6´S	39°46.4´S	14:54:47	14:53:30	14:54:09	36	281	170	03m11s
009° 00.0´E	38°54.6´S	40°26.6´S	39°40.5´S	14:55:54	14:54:35	14:55:15	35	280	169	03m09s
010° 00.0´E	38°48.5´S	40°20.3´S	39°34.2´S	14:56:58	14:55:39	14:56:19	34	279	168	03m06s
011° 00.0´E	38°41.9´S	40°13.4´S	39°27.6´S	14:58:01	14:56:41	14:57:22	33	278	168	03m04s
012° 00.0´E	38°34.9´S	40°06.2´S	39°20.5´S	14:59:02	14:57:41	14:58:22	32	277	167	03m02s
013° 00.0´E	38°27.6´S	39°58.6´S	39°12.9´S	15:00:01	14:58:39	14:59:21	31	276	166	02m59s
014° 00.0´E	38°19.8´S	39°50.5´S	39°05.0´S	15:00:59	14:59:36	15:00:18	30	275	165	02m57s
015° 00.0´E	38°11.7´S	39°42.1´S	38°56.8´S	15:01:54	15:00:31	15:01:13	29	274	165	02m55s
016° 00.0´E	38°03.1´S	39°33.3´S	38°48.1´S	15:02:48	15:01:24	15:02:07	28	273	164	02m53s
017° 00.0´E	37°54.2´S	39°24.1´S	38°39.0´S	15:03:40	15:02:16	15:02:59	27	272	163	02m50s
018° 00.0´E	37°45.0´S	39°14.6´S	38°29.6´S	15:04:30	15:03:06	15:03:49	26	271	162	02m48s
019° 00.0´E	37°35.4´S	39°04.6´S	38°19.9´S	15:05:18	15:03:54	15:04:37	25	270	162	02m46s
020° 00.0´E	37°25.4´S	38°54.4´S	38°09.7´S	15:06:05	15:04:41	15:05:23	25	270	161	02m44s
021° 00.0´E	37°15.1´S	38°43.8´S	37°59.3´S	15:06:49	15:05:26	15:06:08	24	269	160	02m41s
022° 00.0´E	37°04.5´S	38°32.8´S	37°48.5´S	15:07:32	15:06:09	15:06:51	23	268	159	02m39s
023° 00.0´E	36°53.5´S	38°21.5´S	37°37.4´S	15:08:13	15:06:50	15:07:32	22	267	158	02m37s
024° 00.0´E	36°42.3´S	38°09.9´S	37°25.9´S	15:08:52	15:07:30	15:08:12	21	267	158	02m35s
025° 00.0´E	36°30.7´S	37°58.0´S	37°14.2´S	15:09:30	15:08:08	15:08:50	20	266	157	02m33s
026° 00.0´E	36°18.8´S	37°45.8´S	37°02.1´S	15:10:06	15:08:44	15:09:26	19	265	156	02m31s
027° 00.0´E	36°06.7´S	37°33.3´S	36°49.8´S	15:10:40	15:09:19	15:10:60	18	264	155	02m29s
028° 00.0´E	35°54.2´S	37°20.5´S	36°37.2´S	15:11:12	15:09:52	15:10:32	17	264	154	02m27s
029° 00.0´E	35°41.5´S	37°07.4´S	36°24.3´S	15:11:42	15:10:23	15:11:03	16	263	153	02m25s
030° 00.0´E	35°28.6´S	36°54.0´S	36°11.1´S	15:12:11	15:10:53	15:11:32	15	262	152	02m23s
031° 00.0´E	35°15.3´S	36°40.4´S	35°57.7´S	15:12:37	15:11:20	15:12:60	14	262	151	02m21s
032° 00.0´E	35°01.9´S	36°26.5´S	35°44.0´S	15:13:02	15:11:47	15:12:25	13	261	151	02m19s
033° 00.0´E	34°48.2´S	36°12.4´S	35°30.1´S	15:13:26	15:12:11	15:12:49	12	260	150	02m17s
034° 00.0´E	34°34.2´S	35°58.0´S	35°15.9´S	15:13:47	15:12:34	15:13:11	12	260	149	02m15s
035° 00.0´E	34°20.1´S	35°43.4´S	35°01.6´S	15:14:07	15:12:55	15:13:32	11	259	148	02m13s
036° 00.0´E	34°05.7´S	35°28.6´S	34°46.9´S	15:14:25	15:13:14	15:13:50	10	258	147	02m11s
037° 00.0´E	33°50.6´S	35°13.5´S	34°32.1´S	15:14:42	15:13:32	15:14:07	9	258	146	02m09s
038° 00.0´E	33°36.4´S	34°58.2´S	34°17.1´S	15:14:56	15:13:48	15:14:23	8	257	145	02m07s
039° 00.0´E	33°21.4´S	34°42.8´S	34°01.9´S	15:15:09	15:14:02	15:14:36	7	256	144	02m06s
040° 00.0´E	33°06.2´S	34°27.2´S	33°46.5´S	15:15:21	15:14:14	15:14:48	6	256	143	02m04s
041° 00.0´E	32°50.8´S	34°11.4´S	33°30.9´S	15:15:30	15:14:25	15:14:58	5	255	142	02m02s
042° 00.0´E	32°35.4´S	33°55.4´S	33°15.2´S	15:15:38	15:14:34	15:15:07	4	255	141	02m00s
043° 00.0´E	32°19.7´S	33°39.2´S	32°59.3´S	15:15:44	15:14:42	15:15:14	3	254	140	01m59s
044° 00.0´E	32°03.9´S	33°22.9´S	32°43.2´S	15:15:49	15:14:48	15:15:19	2	254	139	01m57s
045° 00.0´E	31°47.9´S	33°06.4´S	32°27.0´S	15:15:52	15:14:53	15:15:23	2	253	138	01m56s
046° 00.0´E	31°31.8´S	32°49.7´S	32°10.6´S	15:15:53	15:15:07	15:15:25	1	252	137	01m54s

Table 8a
CIRCUMSTANCES AT MAXIMUM ECLIPSE ON 3 NOVEMBER 1994
FOR ARGENTINA, BOLIVIA AND CHILE

Location Name	Latitude	Longitude	Elev. m	U.T. of Maximum Eclipse h m s	Umbral Duration	Path Width km	Sun Alt °	Sun Azm °	P °	V °	Eclipse Mag.	Eclipse Obs.
ARGENTINA												
Avellaneda	34°39.0´S	058°23.0´W	—	12:56:53.9			49	72	23	149	0.761	0.708
Bahia Blanca	38°43.0´S	062°17.0´W	31	12:58:45.6			45	71	22	152	0.622	0.538
Buenos Aires	34°36.0´S	058°27.0´W	29	12:56:43.9			49	72	23	149	0.761	0.709
Concordia	31°24.0´S	058°02.0´W	—	12:52:05.8			49	76	24	144	0.843	0.812
Cordoba	31°34.0´S	064°11.0´W	—	12:45:02.3			43	81	23	142	0.761	0.709
Corrientes	27°28.0´S	058°50.0´W	—	12:44:42.8			48	82	24	138	0.929	0.922
Godoy Cruz	32°55.0´S	068°50.0´W	—	12:42:36.9			38	83	22	142	0.675	0.602
La Plata	34°55.0´S	057°57.0´W	—	12:57:53.3			50	71	23	149	0.760	0.707
La Quiaca	22°06.0´S	065°37.0´W	—	12:28:48.0			38	92	23	130	0.973	0.975
Mar del Plata	38°00.0´S	057°33.0´W	—	13:03:20.2			50	66	23	154	0.692	0.624
Mendoza	32°53.0´S	068°49.0´W	861	12:42:34.1			38	84	22	142	0.676	0.603
Paraná	31°44.0´S	060°32.0´W	—	12:49:28.9			47	78	23	144	0.803	0.762
Posadas	27°23.0´S	055°53.0´W	—	12:48:31.6			51	80	24	139	0.971	0.973
Resistencia	27°27.0´S	058°59.0´W	—	12:44:29.7			48	82	24	138	0.928	0.920
Río Cuarto	33°08.0´S	064°21.0´W	—	12:47:24.0			43	80	23	144	0.723	0.661
Rosario	32°58.0´S	060°42.0´W	—	12:51:16.9			47	76	23	145	0.772	0.722
Salta	24°47.0´S	065°25.0´W	—	12:33:02.4			40	89	23	133	0.909	0.896
San Isidro	34°27.0´S	058°30.0´W	—	12:56:25.5			49	72	23	148	0.764	0.713
San Juan	31°32.0´S	068°31.0´W	—	12:40:40.9			38	85	22	141	0.710	0.645
San Miguel de Tuc.	26°49.0´S	065°13.0´W	—	12:36:23.4			40	87	23	136	0.862	0.836
San Salvador de Ju.	24°11.0´S	065°18.0´W	—	12:32:14.3			39	90	23	132	0.925	0.917
Santa Fe	31°38.0´S	060°42.0´W	—	12:49:07.1			46	78	23	143	0.803	0.762
Santiago del Estero	27°47.0´S	064°16.0´W	—	12:38:53.5			42	86	23	137	0.850	0.822
BOLIVIA												
Charaña	17°36.0´S	069°28.0´W	—	12:19:10.0	02m07.5s	172	32	97	203	304	1.046	1.000
Cochabamba	17°24.0´S	066°09.0´W	—	12:21:35.9			36	96	204	304	0.959	0.959
La Paz	16°30.0´S	068°09.0´W	3937	12:18:39.9			33	98	203	303	0.962	0.962
Oruro	17°59.0´S	067°09.0´W	—	12:21:32.1			35	96	203	305	0.988	0.991
Potosi	19°35.0´S	065°45.0´W	—	12:25:00.8	02m42.5s	177	37	94	204	307	1.047	1.000
Rio Mulatos	19°42.0´S	066°47.0´W	—	12:24:14.8	03m04.9s	177	36	95	23	127	1.047	1.000
Santa Cruz	17°48.0´S	063°10.0´W	—	12:24:58.5			39	95	204	305	0.930	0.923
Sucre	19°06.0´S	065°16.0´W	3066	12:24:44.1			38	95	204	306	0.993	0.996
Tarija	21°31.0´S	064°45.0´W	—	12:28:46.6			39	92	24	129	1.000	1.000
Uyuni	20°28.0´S	066°50.0´W	—	12:25:17.8			36	94	23	128	0.999	1.000
Villa Montes	21°15.0´S	063°30.0´W	—	12:29:38.6	03m18.4s	180	40	92	24	129	1.048	1.000
Yacuiba	22°02.0´S	063°45.0´W	—	12:30:33.2			40	91	24	130	1.000	1.000
CHILE												
Antofagasta	23°39.0´S	070°24.0´W	101	12:26:55.9			34	93	23	131	0.874	0.851
Arica	18°29.0´S	070°20.0´W	31	12:19:42.6	01m50.7s	173	32	97	23	125	1.046	1.000
Calama	22°28.0´S	068°56.0´W	—	12:26:23.1			35	93	23	130	0.921	0.912
Chillán	36°36.0´S	072°07.0´W	—	12:45:46.2			36	82	22	146	0.559	0.463
Concepcion	36°50.0´S	073°03.0´W	—	12:45:24.6			35	83	21	146	0.544	0.446
Copiapó	27°22.0´S	070°20.0´W	—	12:32:33.7			35	90	23	135	0.785	0.739
Coquimbo	29°58.0´S	071°21.0´W	—	12:35:46.0			35	88	22	138	0.713	0.648
Iquique	20°13.0´S	070°10.0´W	—	12:22:12.4			32	95	23	127	0.963	0.963
La Serena	29°54.0´S	071°16.0´W	—	12:35:43.8			35	88	22	138	0.715	0.651
Osorno	40°34.0´S	073°09.0´W	—	12:51:27.2			35	79	21	150	0.467	0.360
Puerto Montt	41°28.0´S	072°57.0´W	—	12:53:05.6			36	78	21	151	0.452	0.343
Punta Arenas	53°09.0´S	070°48.0´W	9	13:13:47.1			36	64	19	165	0.263	0.157
Putre	18°11.0´S	069°33.0´W	1312	12:19:52.1	02m58.9s	173	32	97	23	125	1.046	1.000
Rancagua	34°10.0´S	070°45.0´W	—	12:42:57.1			36	84	22	143	0.626	0.542
San Bernardo	33°36.0´S	070°43.0´W	—	12:42:03.9			36	84	22	143	0.638	0.557
Santiago	33°27.0´S	070°40.0´W	—	12:41:52.0			36	84	22	143	0.642	0.562
Talca	35°26.0´S	071°40.0´W	—	12:44:14.3			36	83	22	145	0.588	0.498
Talcahuano	36°43.0´S	073°07.0´W	—	12:45:10.1			35	83	21	146	0.546	0.448
Temuco	38°44.0´S	072°36.0´W	—	12:48:52.3			36	81	21	148	0.510	0.407
Valdivia	39°48.0´S	073°14.0´W	—	12:50:07.5			35	80	21	149	0.482	0.376
Valparaíso	33°02.0´S	071°38.0´W	44	12:40:23.7			35	85	22	142	0.641	0.560
Viña del Mar	33°02.0´S	071°34.0´W	—	12:40:27.0			35	85	22	142	0.641	0.561

Table 8b
LOCAL CIRCUMSTANCES DURING THE TOTAL SOLAR ECLIPSE OF 3 NOVEMBER 1994
FOR ARGENTINA, BOLIVIA AND CHILE

Location Name	First Contact U.T. h m s	Alt °	P °	V °	Second Contact U.T. h m s	Alt °	P °	V °	Third Contact U.T. h m s	Alt °	P °	V °	Fourth Contact U.T. h m s	Alt °	P °	V °
ARGENTINA																
Avellaneda	11:48:54.0	35	306	68									14:10:59.9	62	97	236
Bahía Blanca	11:55:40.8	33	314	80									14:06:25.1	57	88	228
Buenos Aires	11:48:45.6	35	306	68									14:10:49.0	62	97	236
Concordia	11:42:46.5	35	302	60									14:08:34.6	64	102	235
Córdoba	11:40:01.8	29	306	64									13:56:36.7	57	97	223
Corrientes	11:35:17.5	32	298	51									14:02:20.6	64	108	230
Godoy Cruz	11:41:21.8	25	311	70									13:49:32.5	52	92	216
La Plata	11:49:41.7	36	307	69									14:12:07.0	63	97	237
La Quiaca	11:23:32.4	23	295	42									13:42:28.7	55	110	217
Mar del Plata	11:56:23.3	37	310	76									14:15:17.9	61	92	239
Mendoza	11:41:17.7	25	311	70									13:49:31.8	52	92	216
Paraná	11:41:58.6	32	304	62									14:03:51.4	61	100	230
Posadas	11:37:13.3	36	295	48									14:08:17.8	68	110	237
Resistencia	11:35:10.2	32	298	51									14:02:01.3	64	108	230
Río Cuarto	11:43:03.2	29	308	68									13:57:51.2	57	95	223
Rosario	11:44:17.2	33	306	65									14:04:45.8	61	98	230
Salta	11:27:38.2	25	298	49									13:46:26.7	56	107	218
San Isidro	11:48:25.7	35	306	68									14:10:35.0	62	97	235
San Juan	11:38:41.0	25	309	67									13:48:44.9	52	94	216
San Miguel de..	11:31:02.1	26	301	53									13:49:21.7	56	104	219
San Salvador ...	11:26:44.7	25	297	47									13:45:51.7	56	108	218
Santa Fe	11:41:42.0	32	304	62									14:03:25.3	61	100	229
Santiago del ...	11:33:03.2	27	301	55									13:52:15.2	58	103	221
BOLIVIA																
Charaña	11:16:30.2	17	290	34	12:18:07.3	32	68	169	12:20:14.8	32	337	78	13:30:06.3	49	114	213
Cochabamba	11:17:21.2	21	288	31									13:34:39.6	53	117	216
La Paz	11:15:35.7	18	288	31									13:30:17.2	50	117	214
Oruro	11:17:38.9	20	290	34									13:34:02.5	52	115	215
Potosí	11:20:06.7	22	291	36	12:23:40.6	37	81	184	12:26:23.1	37	324	67	13:38:37.1	54	114	217
Río Mulatos	11:19:52.6	21	292	37	12:22:42.5	36	123	226	12:25:47.3	36	282	26	13:37:09.0	53	113	216
Santa Cruz	11:19:10.6	24	287	30									13:40:01.1	57	119	218
Sucre	11:19:40.0	22	290	35									13:38:38.0	55	115	217
Tarija	11:23:04.3	24	293	40									13:43:06.8	56	112	218
Uyuni	11:20:51.4	21	293	39									13:38:10.6	53	112	216
Villa Montes	11:23:16.2	25	292	39	12:27:59.7	40	115	220	12:31:18.1	41	291	36	13:44:52.0	58	113	219
Yacuiba	11:24:13.8	25	293	41									13:45:35.0	58	112	219
CHILE																
Antofagasta	11:24:29.4	19	299	50									13:36:49.0	50	104	213
Arica	11:17:21.1	17	292	37	12:18:46.6	31	164	266	12:20:37.3	32	241	344	13:30:08.4	48	112	213
Calama	11:23:01.9	20	297	46									13:37:35.5	51	107	214
Chillán	11:48:40.4	24	317	81									13:47:13.1	48	84	212
Concepción	11:49:07.4	24	318	82									13:45:53.6	47	83	211
Copiapó	11:30:36.8	21	304	58									13:41:17.4	50	99	213
Coquimbo	11:35:10.9	22	308	65									13:42:25.7	49	94	213
Iquique	11:19:36.2	18	295	41									13:32:44.8	49	110	213
La Serena	11:35:04.0	22	308	65									13:42:29.7	49	94	213
Osorno	11:57:40.4	25	323	91									13:48:27.8	46	78	211
Puerto Montt	11:59:49.6	26	324	93									13:49:22.5	46	77	211
Punta Arenas	12:30:11.3	30	336	120									13:57:49.2	42	61	211
Putre	11:17:09.2	18	291	36	12:18:23.0	32	113	215	12:21:21.9	33	291	33	13:30:48.5	49	113	213
Rancagua	11:43:36.5	24	313	74									13:47:25.5	49	89	214
San Bernardo	11:42:26.3	24	313	73									13:46:58.3	50	90	214
Santiago	11:42:08.2	24	312	73									13:46:55.2	50	90	214
Talca	11:46:10.8	24	315	78									13:46:60.0	48	86	213
Talcahuano	11:48:51.7	23	318	82									13:45:41.6	47	83	211
Temuco	11:53:25.8	25	320	86									13:48:03.9	47	81	212
Valdivia	11:55:52.0	25	322	89									13:47:48.3	46	79	211
Valparaíso	11:41:09.7	23	312	72									13:44:56.5	49	90	213
Viña del Mar	11:41:10.2	23	312	72									13:45:03.0	49	90	213

Table 9a
CIRCUMSTANCES AT MAXIMUM ECLIPSE ON 3 NOVEMBER 1994
FOR BRAZIL

Location Name	Latitude	Longitude	Elev. m	U.T. of Maximum Eclipse h m s	Umbral Duration	Path Width km	Sun Alt °	Sun Azm °	P °	V °	Eclipse Mag.	Eclipse Obs.
BRAZIL												
Alagoinhas	12°07.0'S	038°26.0'W	—	12:56:21.3			70	101	206	290	0.400	0.289
Americana	22°45.0'S	047°20.0'W	—	12:54:44.8			61	80	205	315	0.838	0.806
Anapolis	16°20.0'S	048°58.0'W	—	12:41:47.2			57	93	206	302	0.684	0.614
Aracaju	10°55.0'S	037°04.0'W	—	12:57:46.8			71	105	206	285	0.344	0.232
Aracatuba	21°12.0'S	050°25.0'W	—	12:46:57.7			57	85	205	311	0.840	0.809
Araguari	18°38.0'S	048°11.0'W	—	12:46:42.3			59	88	206	306	0.737	0.679
Arapiraca	09°45.0'S	036°39.0'W	—	12:57:01.9			71	108	206	281	0.303	0.193
Ararangua	28°56.0'S	049°29.0'W	—	13:00:52.4	03m50.8s	189	59	70	24	146	1.052	1.000
Araraquara	21°47.0'S	048°10.0'W	—	12:51:43.2			60	83	205	312	0.823	0.788
Bagé	31°20.0'S	054°06.0'W	—	12:57:27.3			54	72	24	146	0.895	0.880
Barra Mansa	22°32.0'S	044°11.0'W	—	13:00:20.6			65	77	205	316	0.786	0.741
Bauru	22°19.0'S	049°04.0'W	—	12:51:00.3			59	82	205	313	0.851	0.823
Belem	01°27.0'S	048°29.0'W	14	12:22:25.1			49	112	206	280	0.252	0.148
Belo Horizonte	19°55.0'S	043°56.0'W	—	12:56:41.1			65	83	205	310	0.709	0.645
Blumenau	26°56.0'S	049°03.0'W	—	12:58:25.1			59	73	205	322	0.975	0.978
Braganca Paulista	22°57.0'S	046°34.0'W	—	12:56:27.9			62	79	205	316	0.832	0.799
Brasilia	15°47.0'S	047°55.0'W	1142	12:42:44.1			58	93	206	301	0.653	0.576
Cachoeiro de Ita.	20°51.0'S	041°06.0'W	—	13:04:01.4			69	77	205	314	0.694	0.626
Campina Grande	07°13.0'S	035°53.0'W	—	12:55:10.0			71	116	206	274	0.216	0.118
Campinas	22°54.0'S	047°05.0'W	—	12:55:26.4			61	79	205	315	0.838	0.807
Campos	21°45.0'S	041°18.0'W	—	13:05:00.7			69	75	205	316	0.723	0.662
Caruaru	08°17.0'S	035°58.0'W	—	12:56:32.0			71	113	206	277	0.249	0.145
Cascavel	04°07.0'S	038°14.0'W	—	12:45:07.5			65	118	207	273	0.162	0.078
Caxias	04°50.0'S	043°21.0'W	—	12:35:22.0			58	111	206	281	0.267	0.161
Caxias do Sul	29°10.0'S	051°11.0'W	—	12:58:27.5			57	72	24	145	0.989	0.993
Chapecó	27°06.0'S	052°36.0'W	—	12:52:54.6	03m52.4s	188	55	77	24	140	1.051	1.000
Colatina	19°32.0'S	040°37.0'W	—	13:02:59.8			70	81	205	311	0.649	0.571
Criciúma	28°40.0'S	049°23.0'W	—	13:00:37.0	04m01.9s	189	59	70	204	325	1.052	1.000
Cuiabá	15°35.0'S	056°05.0'W	—	12:30:00.2			47	96	205	302	0.771	0.721
Curitiba	25°25.0'S	049°15.0'W	—	12:55:39.0			59	77	205	319	0.937	0.932
Divinópolis	20°09.0'S	044°54.0'W	—	12:55:09.4			64	83	205	310	0.730	0.671
Dourados	22°13.0'S	054°48.0'W	—	12:41:49.7			51	87	205	311	0.930	0.924
Duque de Caxias	22°47.0'S	043°18.0'W	—	13:02:29.9			66	75	205	317	0.781	0.734
Feira de Santana	12°15.0'S	038°57.0'W	—	12:55:21.8			69	100	206	291	0.412	0.302
Florianópolis	27°35.0'S	048°34.0'W	31	13:00:17.1			60	71	204	324	0.986	0.989
Fortaleza	03°35.0'S	038°31.0'W	—	12:43:45.4			64	118	207	272	0.151	0.070
Foz do Iguacu	25°33.0'S	054°35.0'W	—	12:47:26.4	03m26.4s	187	52	81	204	317	1.051	1.000
Franca	20°32.0'S	047°24.0'W	—	12:51:05.8			60	84	205	310	0.778	0.730
Goiania	16°40.0'S	049°16.0'W	—	12:41:47.5			57	92	206	303	0.698	0.631
Gov. Valadares	18°51.0'S	041°56.0'W	—	12:59:05.6			68	84	206	308	0.649	0.571
Guarulhos	23°28.0'S	046°32.0'W	—	12:57:21.0			62	78	205	317	0.846	0.817
Ilhéus	14°49.0'S	039°02.0'W	—	12:59:08.6			71	93	206	297	0.489	0.384
Imperatriz	05°32.0'S	047°29.0'W	—	12:28:55.3			53	108	206	285	0.354	0.242
Ipatinga	19°30.0'S	042°32.0'W	—	12:58:52.1			67	83	206	310	0.676	0.605
Itabuna	14°48.0'S	039°16.0'W	—	12:58:35.0			70	93	206	297	0.492	0.388
Itajaí	26°53.0'S	048°39.0'W	—	12:59:01.5			60	73	205	322	0.968	0.970
Itapetininga	23°36.0'S	048°03.0'W	—	12:54:49.5			60	79	205	316	0.871	0.849
Jacareí	23°19.0'S	045°58.0'W	—	12:58:09.9			63	77	205	317	0.834	0.801
Jequié	13°51.0'S	040°05.0'W	—	12:55:16.0			69	96	206	295	0.476	0.371
João Pessoa	07°07.0'S	034°52.0'W	—	12:57:36.8			72	118	206	271	0.197	0.103
Joinvile	26°18.0'S	048°50.0'W	—	12:57:46.6			60	74	205	321	0.955	0.954
Juázeiro do Norte	07°12.0'S	039°20.0'W	—	12:46:57.8			66	111	206	280	0.270	0.164
Juiz de Fora	21°45.0'S	043°20.0'W	—	13:00:47.8			66	78	205	315	0.752	0.698
Jundiaí	23°11.0'S	046°52.0'W	—	12:56:17.2			62	79	205	316	0.843	0.813
Lajes	27°48.0'S	050°19.0'W	—	12:57:40.8	03m37.9s	189	58	73	204	323	1.052	1.000
Limeira	22°34.0'S	047°24.0'W	—	12:54:20.1			61	80	205	314	0.834	0.801
Linhares	19°25.0'S	040°04.0'W	—	13:04:01.4			70	80	205	311	0.638	0.558
Londrina	23°18.0'S	051°09.0'W	—	12:49:06.4			56	82	205	314	0.907	0.895
Macapá	00°02.0'N	051°03.0'W	—	12:17:08.1			45	112	206	280	0.249	0.145

Table 9b
LOCAL CIRCUMSTANCES DURING THE TOTAL SOLAR ECLIPSE OF 3 NOVEMBER 1994 FOR BRAZIL

Location Name	First Contact				Second Contact				Third Contact				Fourth Contact			
	U.T. h m s	Alt °	P °	V °	U.T. h m s	Alt °	P °	V °	U.T. h m s	Alt °	P °	V °	U.T. h m s	Alt °	P °	V °
BRAZIL																
Alagoinhas	11:50:45.1	54	257	347									14:11:44.0	87	151	175
Americana	11:39:32.1	44	283	30									14:19:42.8	79	124	259
Anapolis	11:31:05.4	40	274	13									14:03:07.3	76	134	230
Aracaju	11:55:09.4	56	253	340									14:09:49.9	86	155	162
Aracatuba	11:33:49.1	40	283	28									14:10:12.3	76	124	241
Araguari	11:34:03.9	42	277	19									14:09:49.1	78	130	239
Arapiraca	11:57:21.4	57	250	335									14:05:53.6	85	158	169
Ararangua	11:45:39.5	43	293	49	12:58:56.4	59	130	251	13:02:47.2	59	276	37	14:24:11.2	74	112	261
Araraquara	11:37:18.1	43	282	28									14:16:07.5	78	125	251
Bagé	11:45:27.7	39	299	58									14:16:47.9	68	105	246
Barra Mansa	11:43:53.0	48	280	27									14:26:21.8	82	126	283
Bauru	11:36:46.3	42	284	30									14:15:08.1	77	123	249
Belem	11:34:01.7	38	246	325									13:18:15.6	62	164	227
Belo Horizonte	11:41:44.9	48	276	19									14:21:41.6	84	131	270
Blumenau	11:43:01.6	43	290	43									14:22:31.3	75	116	260
Braganca Paul...	11:40:49.4	45	283	30									14:21:46.4	80	124	264
Brasilia	11:32:14.0	41	273	11									14:03:54.1	78	135	230
Cachoeiro de ...	11:47:33.7	52	275	19									14:29:59.6	84	132	316
Campina Grande	12:03:18.0	59	243	322									13:55:30.7	82	166	186
Campinas	11:40:03.1	44	283	30									14:20:32.1	79	124	261
Campos	11:47:58.6	51	277	22									14:31:23.0	83	130	315
Caruaru	12:01:19.5	58	246	327									14:00:31.0	83	163	177
Cascavel	12:00:48.8	55	238	313									13:37:04.6	75	172	214
Caxias	11:42:49.9	46	247	328									13:36:25.1	72	163	220
Caxias do Sul	11:44:17.3	41	295	50									14:20:41.7	72	110	254
Chapecó	11:39:34.9	39	293	46	12:50:58.5	55	118	234	12:54:50.9	56	288	44	14:14:55.4	72	113	246
Colatina	11:47:29.3	52	272	15									14:28:11.6	86	134	321
Criciúma	11:45:20.2	43	293	48	12:58:36.4	59	111	232	13:02:38.4	59	294	55	14:24:04.7	74	112	261
Cuiabá	11:22:08.4	31	279	18									13:48:05.6	66	129	223
Curitiba	11:40:35.7	42	288	39									14:19:51.5	76	118	256
Divinópolis	11:40:22.3	46	277	20									14:20:03.8	83	130	263
Dourados	11:30:30.4	35	288	34									14:02:51.8	70	119	232
Duque de Caxi...	11:45:33.7	49	280	27									14:28:49.6	82	127	293
Feira de Sant...	11:49:21.7	53	258	348									14:11:13.7	87	150	185
Florianópolis	11:44:33.7	44	291	44									14:24:29.9	75	115	264
Fortaleza	12:01:01.8	54	237	311									13:33:54.9	74	173	217
Foz do Iguacu	11:35:23.9	36	292	43	12:45:44.6	52	88	201	12:49:11.0	53	318	70	14:08:30.0	70	114	238
Franca	11:36:59.6	43	280	24									14:15:23.7	79	128	249
Goiania	11:30:55.1	40	275	14									14:03:17.2	76	133	231
Gov. Valadares	11:44:28.9	50	273	14									14:23:42.7	86	135	288
Guarulhos	11:41:29.2	45	283	31									14:22:45.2	79	123	266
Ilhéus	11:48:58.1	54	263	357									14:19:15.2	90	145	161
Imperatriz	11:32:16.8	40	254	338									13:34:35.4	68	156	222
Ipatinga	11:43:51.0	49	274	16									14:23:52.2	85	133	285
Itabuna	11:48:24.9	53	263	358									14:18:43.6	89	145	194
Itajaí	11:43:24.8	43	290	42									14:23:20.8	76	116	262
Itapetininga	11:39:38.0	43	285	33									14:19:36.1	78	122	258
Jacareí	11:42:06.1	46	283	31									14:23:45.8	80	124	270
Jequié	11:46:25.4	52	262	355									14:14:12.8	87	146	207
João Pessoa	12:07:20.4	61	241	318									13:56:11.0	82	167	179
Joinvile	11:42:20.6	43	289	41									14:22:06.2	76	117	260
Juázeiro do N..	11:51:42.7	53	248	330									13:51:09.6	79	162	205
Juiz de Fora	11:44:26.8	49	278	24									14:26:47.1	83	128	288
Jundiaí	11:40:41.2	45	283	31									14:21:31.0	79	123	263
Lajes	11:42:57.8	42	292	46	12:55:53.3	58	88	207	12:59:31.2	58	318	76	14:20:52.8	74	113	256
Limeira	11:39:13.8	44	283	30									14:19:14.3	79	124	258
Linhares	11:48:30.2	53	272	14									14:29:09.1	86	135	332
Londrina	11:35:34.2	39	287	34									14:12:19.3	74	120	244
Macapá	11:30:15.5	34	245	324									13:10:56.6	57	164	229

Table 9a
CIRCUMSTANCES AT MAXIMUM ECLIPSE ON 3 NOVEMBER 1994
FOR BRAZIL

Location Name	Latitude	Longitude	Elev. m	U.T. of Maximum Eclipse h m s	Umbral Duration	Path Width km	Sun Alt °	Sun Azm °	P °	V °	Eclipse Mag.	Eclipse Obs.
BRAZIL												
Maceió	09°40.0'S	035°43.0'W	—	12:59:14.2			73	110	206	280	0.286	0.178
Manaus	03°08.0'S	060°01.0'W	47	12:10:04.8			36	106	205	287	0.477	0.371
Marabá	05°21.0'S	049°07.0'W	—	12:26:04.5			51	107	206	286	0.375	0.263
Marília	22°13.0'S	049°56.0'W	—	12:49:22.7			58	83	205	313	0.861	0.836
Maringá	23°25.0'S	051°55.0'W	—	12:48:04.3			55	83	205	314	0.921	0.912
Montes Claros	16°43.0'S	043°52.0'W	—	12:51:46.0			64	90	206	303	0.617	0.533
Mossoró	05°11.0'S	037°20.0'W	—	12:48:42.1			67	117	206	273	0.179	0.090
Natal	05°47.0'S	035°13.0'W	17	12:54:46.2			71	120	206	270	0.163	0.078
Niterói	22°53.0'S	043°07.0'W	—	13:03:01.6			67	75	205	318	0.781	0.734
Nova Friburgo	22°16.0'S	042°32.0'W	—	13:03:14.6			67	75	205	317	0.755	0.702
Nova Iguacu	22°45.0'S	043°27.0'W	—	13:02:08.7			66	75	205	317	0.782	0.736
Parnaíba	02°54.0'S	041°47.0'W	—	12:35:55.0			59	115	207	276	0.185	0.094
Passo Fundo	28°15.0'S	052°24.0'W	—	12:55:04.1			55	75	24	142	0.996	0.998
Patos de Minas	18°35.0'S	046°32.0'W	—	12:49:35.4			61	88	206	307	0.710	0.646
Pelotas	31°46.0'S	052°20.0'W	—	13:00:48.2			56	69	24	148	0.907	0.895
Petrolina	09°24.0'S	040°30.0'W	—	12:47:36.0			65	106	206	286	0.353	0.242
Petrópolis	22°31.0'S	043°10.0'W	—	13:02:20.8			67	76	205	317	0.771	0.722
Piracicaba	22°43.0'S	047°38.0'W	—	12:54:09.3			61	80	205	315	0.841	0.811
Poços de Caldas	21°48.0'S	046°34.0'W	—	12:54:37.9			62	81	205	313	0.800	0.759
Ponta Grossa	25°05.0'S	050°09.0'W	—	12:53:36.1			58	78	205	318	0.941	0.937
Porto Alegre	30°04.0'S	051°11.0'W	11	12:59:53.8			57	70	24	146	0.966	0.967
Porto Velho	08°46.0'S	063°54.0'W	—	12:12:38.9			34	102	204	294	0.693	0.624
Pres. Prudente	22°07.0'S	051°22.0'W	—	12:46:52.5			56	84	205	312	0.879	0.858
Recife	08°09.0'S	034°59.0'W	32	12:58:50.4			73	115	206	274	0.230	0.129
Ribeirão Prêto	21°10.0'S	047°48.0'W	—	12:51:23.2			60	83	205	311	0.801	0.760
Rio Branco	09°58.0'S	067°48.0'W	—	12:10:56.7			30	102	203	296	0.779	0.730
Rio Claro	22°24.0'S	047°33.0'W	—	12:53:47.9			61	81	205	314	0.831	0.798
Rio Grande	32°02.0'S	052°05.0'W	—	13:01:36.7			56	68	24	149	0.904	0.891
Rio de Janeiro	22°54.0'S	043°14.0'W	66	13:02:49.0			67	75	205	318	0.783	0.737
Rondonópolis	16°28.0'S	054°38.0'W	—	12:33:15.5			49	94	205	303	0.773	0.725
Salvador	12°59.0'S	038°27.0'W	51	12:57:39.0			71	99	206	292	0.426	0.316
Santa Cruz do Sul	29°43.0'S	052°26.0'W	—	12:57:22.4			56	72	24	145	0.958	0.958
Santa Maria	29°41.0'S	053°48.0'W	—	12:55:14.8			54	74	24	144	0.941	0.937
Santarém	02°26.0'S	054°42.0'W	—	12:15:03.2			42	108	205	285	0.378	0.266
Santo Andre	23°40.0'S	046°31.0'W	—	12:57:42.0			62	77	205	317	0.851	0.823
Santo Angelo	28°18.0'S	054°16.0'W	—	12:52:19.6			53	77	24	141	0.970	0.972
Santos	23°57.0'S	046°20.0'W	—	12:58:29.4			63	76	205	318	0.856	0.830
São Caetano do Sul	23°36.0'S	046°34.0'W	—	12:57:30.1			62	77	205	317	0.850	0.822
São Carlos	22°01.0'S	047°54.0'W	—	12:52:33.8			60	82	205	313	0.826	0.791
São Gonçalo	22°51.0'S	043°04.0'W	—	13:03:04.5			67	75	205	318	0.779	0.732
São José Rio Preto	20°48.0'S	049°23.0'W	—	12:48:03.2			58	85	205	310	0.814	0.776
São José dos Campos	22°10.0'S	045°06.0'W	—	12:57:58.7			64	79	205	315	0.789	0.745
São Luis	02°31.0'S	044°16.0'W	—	12:30:42.3			56	113	206	278	0.214	0.117
São Paulo	23°32.0'S	046°37.0'W	862	12:57:17.8			62	77	205	317	0.849	0.821
Sete Lagoas	19°27.0'S	044°14.0'W	—	12:55:21.0			65	84	206	309	0.700	0.634
Sobral	03°42.0'S	040°21.0'W	—	12:39:54.3			62	115	207	275	0.185	0.094
Sorocaba	23°29.0'S	047°27.0'W	—	12:55:42.5			61	79	205	316	0.859	0.834
Taubaté	23°02.0'S	045°33.0'W	—	12:58:29.9			63	77	205	316	0.820	0.784
Teófilo Otoni	17°51.0'S	041°30.0'W	—	12:58:26.1			68	86	206	306	0.614	0.529
Teresina	05°05.0'S	042°49.0'W	—	12:36:43.8			59	111	207	281	0.265	0.159
Teresópolis	22°26.0'S	042°59.0'W	—	13:02:35.2			67	76	205	317	0.766	0.716
Uberaba	19°45.0'S	047°55.0'W	—	12:48:56.0			60	86	205	309	0.764	0.713
Uberlândia	18°56.0'S	048°18.0'W	—	12:46:58.3			59	88	205	307	0.747	0.691
Uruguaiana	29°45.0'S	057°05.0'W	—	12:50:41.6			50	77	24	142	0.896	0.880
Vicente de Carvalho	23°56.0'S	046°19.0'W	—	12:58:29.7			63	76	205	318	0.856	0.829
Vitória	20°19.0'S	040°21.0'W	—	13:04:48.8			70	78	205	313	0.668	0.594
Vitória Conquista	14°51.0'S	040°51.0'W	—	12:55:08.0			68	94	206	298	0.517	0.416
Vitória de S. Antão	08°07.0'S	035°18.0'W	—	12:57:58.5			72	115	206	275	0.234	0.133
Volta Redonda	22°32.0'S	044°07.0'W	—	13:00:28.5			65	77	205	316	0.785	0.740

Table 9b
LOCAL CIRCUMSTANCES DURING THE TOTAL SOLAR ECLIPSE OF 3 NOVEMBER 1994
FOR BRAZIL

Location Name	First Contact U.T. h m s	Alt °	P °	V °	Second Contact U.T. h m s	Alt °	P °	V °	Third Contact U.T. h m s	Alt °	P °	V °	Fourth Contact U.T. h m s	Alt °	P °	V °
BRAZIL																
Maceió	12:00:26.7	59	249	332									14:07:02.5	85	160	158
Manaus	11:14:12.2	23	261	348									13:14:01.8	51	147	222
Marabá	11:29:06.7	37	255	340									13:32:02.9	66	154	222
Marília	11:35:35.6	41	284	30									14:13:04.8	76	123	245
Maringá	11:34:54.2	39	287	35									14:10:52.3	74	119	241
Montes Claros	11:39:33.5	47	271	9									14:14:25.7	84	137	242
Mossoró	12:01:50.5	56	239	316									13:43:32.2	78	170	207
Natal	12:08:54.4	60	238	313									13:48:32.9	80	171	194
Niterói	11:45:58.0	49	280	27									14:29:25.6	82	126	295
Nova Friburgo	11:46:18.2	50	278	25									14:29:35.6	83	128	300
Nova Iguacu	11:45:17.0	49	280	27									14:28:25.4	82	127	291
Parnaíba	11:50:38.4	49	240	317									13:28:42.3	70	170	223
Passo Fundo	11:41:36.7	40	294	48									14:16:52.7	71	111	248
Patos de Minas	11:36:26.4	44	276	17									14:13:10.9	81	132	244
Pelotas	11:47:44.3	41	299	58									14:21:04.9	70	106	252
Petrolina	11:46:29.9	51	254	340									13:58:21.1	81	155	205
Petrópolis	11:45:30.0	49	279	26									14:28:37.9	82	127	293
Piracicaba	11:39:05.6	43	283	30									14:18:59.4	79	124	257
Pocos de Cald..	11:39:30.9	45	281	27									14:19:40.1	80	126	260
Ponta Grossa	11:39:05.9	41	288	38									14:17:21.5	75	118	252
Porto Alegre	11:45:48.3	41	296	53									14:21:45.6	72	109	255
Porto Velho	11:11:39.6	20	273	8									13:22:29.8	51	133	218
Pres. Prudente	11:33:49.5	39	285	31									14:09:50.3	74	122	240
Recife	12:05:01.9	60	244	324									14:01:14.1	83	164	169
Ribeirão Prêto	11:37:06.4	43	281	26									14:15:45.7	79	126	250
Rio Branco	11:10:16.9	16	278	14									13:20:05.1	47	127	216
Rio Claro	11:38:49.9	43	283	29									14:18:36.4	79	124	256
Rio Grande	11:48:26.7	41	299	58									14:21:54.0	70	105	253
Rio de Janeiro	11:45:47.6	49	280	27									14:29:11.5	82	126	294
Rondonópolis	11:24:19.8	33	279	19									13:52:31.7	68	128	225
Salvador	11:50:29.2	54	259	350									14:14:38.5	88	149	166
Santa Cruz do..	11:44:02.0	40	296	52									14:18:36.4	71	109	250
Santa Maria	11:42:46.7	38	297	53									14:15:35.9	69	108	245
Santarém	11:21:12.7	29	255	339									13:16:55.3	57	154	225
Santo Andre	11:41:45.2	45	284	32									14:23:08.1	79	123	267
Santo Angelo	11:40:02.2	38	296	50									14:12:54.4	70	110	242
Santos	11:42:21.3	45	284	33									14:24:01.7	79	122	269
São Caetano d..	11:41:36.2	45	284	32									14:22:54.5	79	123	267
São Carlos	11:37:55.3	43	282	28									14:17:08.3	79	125	253
São Goncalo	11:46:00.8	49	280	27									14:29:28.7	82	127	296
São José Rio ...	11:34:38.0	41	282	26									14:11:39.1	77	126	243
São José dos ...	11:42:04.3	46	280	27									14:23:37.1	81	126	272
São Luis	11:43:33.2	45	243	321									13:25:27.4	68	167	225
São Paulo	11:41:26.5	45	283	32									14:22:40.9	79	123	266
Sete Lagoas	11:40:51.0	47	275	18									14:20:01.7	83	132	263
Sobral	11:53:55.2	51	240	317									13:33:36.5	73	170	219
Sorocaba	11:40:16.1	44	284	32									14:20:44.2	79	122	261
Taubaté	11:42:22.2	46	282	29									14:24:11.9	81	125	272
Teófilo Otoni	11:44:41.2	50	270	10									14:22:16.1	87	137	283
Teresina	11:44:00.4	47	247	328									13:38:01.1	73	163	218
Teresópolis	11:45:43.0	49	279	26									14:28:53.4	83	127	295
Uberaba	11:35:29.2	42	279	22									14:12:42.6	79	129	244
Uberlândia	11:34:10.6	42	278	20									14:10:12.5	78	130	240
Uruguaiana	11:40:23.3	35	299	55									14:08:42.9	66	105	236
Vicente de Ca..	11:42:21.5	45	284	33									14:24:02.3	79	122	270
Vitória	11:48:34.4	52	273	17									14:30:32.3	85	133	327
Vitória Conqu..	11:44:49.6	51	265	360									14:15:38.8	87	143	226
Vitória de S....	12:03:56.0	60	244	324									14:00:38.8	83	164	172
Volta Redonda	11:43:59.3	48	280	27									14:26:30.9	82	126	283

Table 10a
CIRCUMSTANCES AT MAXIMUM ECLIPSE ON 3 NOVEMBER 1994
FOR COLOMBIA, EQUADOR AND VENEZUELA

Location Name	Latitude	Longitude	Elev. m	U.T. of Maximum Eclipse h m s	Umbral Duration	Path Width km	Sun Alt °	Sun Azm °	P °	V °	Eclipse Mag.	Eclipse Obs.
COLOMBIA												
Armenia	04°31.0'N	075°41.0'W	–	11:55:03.9			15	107	202	283	0.452	0.343
Barrancabermeja	07°03.0'N	073°52.0'W	–	11:54:19.3			16	108	202	280	0.354	0.242
Barranquilla	10°59.0'N	074°48.0'W	–	11:52:35.5			14	109	201	276	0.244	0.141
Bogota	04°36.0'N	074°05.0'W	2741	11:55:26.0			17	107	202	282	0.431	0.321
Bucaramanga	07°08.0'N	073°09.0'W	–	11:54:30.3			17	108	202	280	0.343	0.231
Buenaventura	03°53.0'N	077°04.0'W	–	11:55:04.1			14	107	202	283	0.487	0.380
Buga	03°54.0'N	076°17.0'W	–	11:55:14.7			15	107	202	283	0.477	0.370
Cali	03°27.0'N	076°31.0'W	–	11:55:26.3			15	107	201	284	0.494	0.388
Cartagena	10°25.0'N	075°32.0'W	–	11:52:35.9			13	108	201	277	0.269	0.163
Cartago	04°45.0'N	075°55.0'W	–	11:54:52.9			15	107	202	282	0.447	0.338
Cúcuta	07°54.0'N	072°31.0'W	–	11:54:22.1			17	109	202	279	0.312	0.202
Girardot	04°18.0'N	074°48.0'W	–	11:55:25.7			16	107	202	283	0.448	0.339
Ibagué	04°27.0'N	075°14.0'W	–	11:55:13.4			16	107	202	283	0.449	0.339
Itaguí	06°10.0'N	075°36.0'W	–	11:54:15.5			15	107	202	281	0.401	0.289
Manizales	05°05.0'N	075°32.0'W	–	11:54:48.5			15	107	202	282	0.433	0.322
Medellín	06°15.0'N	075°35.0'W	–	11:54:13.3			15	107	202	281	0.398	0.286
Montería	08°46.0'N	075°53.0'W	–	11:53:05.5			13	108	201	278	0.324	0.213
Neiva	02°56.0'N	075°18.0'W	–	11:56:03.8			16	107	202	284	0.495	0.390
Palmira	03°32.0'N	076°16.0'W	–	11:55:27.2			15	107	202	284	0.488	0.382
Pasto	01°13.0'N	077°17.0'W	–	11:56:35.3			15	106	201	286	0.569	0.475
Pereira	04°49.0'N	075°43.0'W	–	11:54:53.9			15	107	202	282	0.443	0.333
Popayán	02°27.0'N	076°36.0'W	–	11:55:59.8			15	106	202	285	0.525	0.423
Santa Marta	11°15.0'N	074°13.0'W	–	11:52:39.2			14	109	202	276	0.229	0.128
Sincelejo	09°18.0'N	075°24.0'W	–	11:53:00.3			14	108	201	278	0.302	0.192
Tuluá	04°06.0'N	076°11.0'W	–	11:55:09.7			15	107	202	283	0.470	0.363
Valledupar	10°29.0'N	073°15.0'W	–	11:53:09.2			15	109	202	276	0.242	0.139
Villavicencio	04°09.0'N	073°37.0'W	–	11:55:52.9			18	107	202	283	0.439	0.329
ECUADOR												
Ambato	01°15.0'S	078°37.0'W	–	11:57:55.8			15	105	201	289	0.657	0.579
Cuenca	02°53.0'S	078°59.0'W	–	11:59:02.6			15	105	201	290	0.709	0.643
Esmeraldas	00°59.0'N	079°42.0'W	–	11:56:13.9			13	106	201	287	0.602	0.513
Guayaquil	02°10.0'S	079°50.0'W	7	11:58:19.6			14	105	201	290	0.697	0.628
Machala	03°16.0'S	079°58.0'W	–	11:59:07.0			14	105	201	291	0.731	0.670
Manta	00°57.0'N	080°44.0'W	–	11:57:18.7			13	105	201	289	0.671	0.596
Milagro	02°07.0'S	079°36.0'W	–	11:58:20.2			14	105	201	290	0.693	0.623
Portoviejo	01°03.0'S	080°27.0'W	–	11:57:25.5			13	105	201	289	0.671	0.596
Quito	00°13.0'S	078°30.0'W	3026	11:57:11.6			14	106	201	288	0.626	0.541
Riobamba	01°40.0'S	078°38.0'W	–	11:58:13.4			15	105	201	289	0.670	0.594
VENEZUELA												
Acarigua	09°33.0'N	069°12.0'W	–	11:54:56.5			20	110	202	276	0.222	0.123
Barcelona	10°08.0'N	064°42.0'W	–	11:56:59.3			24	112	203	275	0.147	0.067
Barinas	08°38.0'N	070°12.0'W	–	11:54:54.3			19	109	202	277	0.262	0.156
Barquismeto	10°04.0'N	069°19.0'W	–	11:54:41.4			19	110	202	276	0.208	0.111
Cabimas	10°23.0'N	071°28.0'W	–	11:53:45.1			17	109	202	276	0.224	0.124
Caracas	10°30.0'N	066°56.0'W	1121	11:55:36.2			22	111	203	275	0.165	0.079
Ciudad Bolívar	08°08.0'N	063°33.0'W	–	11:58:38.4			26	111	203	276	0.192	0.099
Ciudad Guayana	08°22.0'N	062°40.0'W	–	11:59:06.7			27	112	204	276	0.173	0.085
Ciudad Ojeda	10°12.0'N	071°19.0'W	–	11:53:52.1			17	109	202	276	0.228	0.127
Coro	11°25.0'N	069°41.0'W	–	11:54:04.0			19	110	202	274	0.171	0.083
Cumaná	10°28.0'N	064°10.0'W	–	11:57:10.1			25	112	203	274	0.130	0.056
Maracaibo	10°40.0'N	071°37.0'W	7	11:53:36.2			17	109	202	276	0.217	0.119
Maracay	10°15.0'N	067°36.0'W	–	11:55:23.6			21	111	203	275	0.181	0.091
Maturín	09°45.0'N	063°11.0'W	–	11:58:05.7			26	112	203	274	0.139	0.061
Mérida	08°36.0'N	071°08.0'W	–	11:54:33.4			18	109	202	278	0.275	0.167
Puerto Cabello	10°28.0'N	068°01.0'W	–	11:55:06.9			21	111	203	275	0.179	0.090
Puerto la Cruz	10°13.0'N	064°38.0'W	–	11:56:59.6			24	112	203	274	0.144	0.065
San Cristóbal	07°46.0'N	072°14.0'W	–	11:54:31.4			17	109	202	279	0.313	0.202
Valencia	10°11.0'N	068°00.0'W	–	11:55:13.8			21	111	203	275	0.188	0.096
Valera	09°19.0'N	070°37.0'W	–	11:54:27.5			18	109	202	277	0.246	0.143

Table 10b
LOCAL CIRCUMSTANCES DURING THE TOTAL SOLAR ECLIPSE OF 3 NOVEMBER 1994
FOR COLOMBIA, EQUADOR AND VENEZEULA

Location Name	First Contact U.T. h m s	Alt °	P °	V °	Second Contact U.T. h m s	Alt °	P °	V °	Third Contact U.T. h m s	Alt °	P °	V °	Fourth Contact U.T. h m s	Alt °	P °	V °
COLOMBIA																
Armenia	11:07:19.6	4	257	341									12:47:55.4	28	146	223
Barrancaberme...	11:10:13.1	6	250	331									12:42:53.8	27	153	227
Barranquilla	11:15:03.1	5	241	318									12:33:19.8	23	161	232
Bogota	11:07:55.7	6	255	339									12:48:06.6	29	148	223
Bucaramanga	11:10:42.2	7	249	330									12:42:45.9	28	154	227
Buenaventura	11:06:33.7	3	259	344									12:48:46.7	27	144	221
Buga	11:06:46.7	4	258	343									12:48:57.3	28	144	222
Cali	11:06:28.9	3	259	345									12:49:43.9	28	143	221
Cartagena	11:13:38.6	4	243	321									12:34:57.1	23	159	231
Cartago	11:07:24.1	4	256	340									12:47:24.7	27	146	223
Cúcuta	11:11:59.9	7	247	327									12:40:57.6	28	156	229
Girardot	11:07:29.8	5	257	341									12:48:34.6	29	146	223
Ibagué	11:07:26.3	5	256	341									12:48:10.0	28	146	223
Itaguí	11:08:35.1	4	253	336									12:44:36.7	27	149	225
Manizales	11:07:45.7	4	255	339									12:46:50.0	28	147	223
Medellín	11:08:39.8	4	253	335									12:44:26.5	27	150	225
Montería	11:11:09.5	4	247	327									12:38:56.5	24	155	228
Neiva	11:06:34.6	5	260	345									12:51:04.9	29	143	221
Palmira	11:06:35.3	4	259	344									12:49:38.9	28	144	221
Pasto	11:05:30.8	3	264	352									12:53:26.1	29	138	218
Pereira	11:07:30.8	4	256	340									12:47:19.5	28	147	223
Popayán	11:06:01.6	3	261	348									12:51:31.7	29	141	220
Santa Marta	11:15:57.3	6	240	317									12:32:27.1	23	163	233
Sincelejo	11:12:05.8	4	246	325									12:37:40.0	24	157	229
Tuluá	11:06:55.1	4	258	343									12:48:36.1	28	145	222
Valledupar	11:15:19.5	7	241	318									12:34:19.8	25	162	233
Villavicencio	11:07:52.2	6	256	340									12:49:12.4	30	147	223
ECUADOR																
Ambato	11:05:05.5	2	269	360									12:56:51.9	29	133	216
Cuenca	11:05:13.4	2	272	5									12:59:09.6	30	130	214
Esmeraldas	11:05:10.6	1	266	355									12:52:54.4	26	136	217
Guayaquil	11:05:06.1	1	271	3									12:57:39.7	28	130	214
Machala	11:05:16.9	2	273	6									12:59:11.7	29	128	214
Manta	11:05:00.7	0	270	1									12:55:27.6	27	132	215
Milagro	11:05:06.1	2	271	3									12:57:42.2	28	131	215
Portoviejo	11:05:00.9	1	270	1									12:55:43.7	27	132	215
Quito	11:05:04.7	2	267	357									12:55:14.8	28	135	217
Riobamba	11:05:06.4	2	270	1									12:57:30.1	29	132	215
VENEZUELA																
Acarigua	11:17:18.8	11	240	317									12:36:10.2	29	164	234
Barcelona	11:24:32.8	17	233	308									12:32:34.2	32	172	239
Barinas	11:14:47.7	10	243	322									12:38:58.6	29	161	231
Barquisimeto	11:18:11.0	11	239	315									12:34:36.1	29	165	235
Cabimas	11:16:39.0	9	240	317									12:34:12.0	26	164	234
Caracas	11:22:03.4	14	235	310									12:32:16.4	30	170	238
Ciudad Bolívar	11:21:37.7	18	238	314									12:39:35.8	36	168	236
Ciudad Guayana	11:23:31.3	19	236	312									12:38:29.5	36	170	237
Ciudad Ojeda	11:16:26.9	9	240	317									12:34:42.0	27	163	233
Coro	11:20:42.1	11	235	310									12:30:19.2	27	169	237
Cumaná	11:26:23.7	18	232	305									12:30:53.5	32	174	241
Maracaibo	11:17:02.9	8	239	316									12:33:24.3	26	164	234
Maracay	11:20:35.9	13	236	312									12:33:26.9	30	168	237
Maturín	11:26:07.6	19	233	307									12:33:15.9	34	173	240
Mérida	11:13:57.7	9	244	323									12:39:08.0	29	160	231
Puerto Cabello	11:20:33.3	13	236	312									12:32:51.9	29	168	237
Puerto la Cruz	11:24:52.8	17	233	307									12:32:12.2	32	172	240
San Cristóbal	11:12:01.1	8	247	327									12:41:17.7	28	156	229
Valencia	11:19:56.8	13	237	313									12:33:49.1	30	167	236
Valera	11:15:31.9	9	242	320									12:37:06.4	28	162	232

Table 11a
CIRCUMSTANCES AT MAXIMUM ECLIPSE ON 3 NOVEMBER 1994
FOR PARAGUAY, PERU AND URUGUAY

Location Name	Latitude	Longitude	Elev. m	U.T. of Maximum Eclipse h m s	Umbral Duration	Path Width km	Sun Alt °	Sun Azm °	P °	V °	Eclipse Mag.	Eclipse Obs.
PARAGUAY												
Asuncion	25°16.0′S	057°40.0′W	150	12:42:42.6	00m40.7s	186	49	84	24	135	1.050	1.000
Caaguazu	25°00.0′S	055°45.0′W	—	12:44:53.6	03m29.9s	186	51	83	204	315	1.050	1.000
Concepcion	23°25.0′S	057°17.0′W	—	12:40:17.6			49	86	204	313	0.997	0.999
Encarnación	27°20.0′S	055°54.0′W	—	12:48:25.4			51	80	24	139	0.972	0.974
Fernando d.l. Mora	25°19.0′S	057°36.0′W	—	12:42:52.8	00m20.0s	186	49	84	24	135	1.050	1.000
Marsiscal	22°02.0′S	060°38.0′W	—	12:33:58.5	02m22.3s	182	44	90	204	310	1.049	1.000
Rosario	24°27.0′S	057°03.0′W	—	12:42:14.1	03m35.3s	186	49	85	204	314	1.050	1.000
Villarrica	25°45.0′S	056°26.0′W	—	12:45:08.6	02m14.4s	186	50	83	24	136	1.050	1.000
PERU												
Arequipa	16°24.0′S	071°33.0′W	2776	12:16:05.9	00m57.6s	169	29	98	203	303	1.045	1.000
Ayacucho	13°07.0′S	074°13.0′W	—	12:10:35.7			25	101	202	300	0.945	0.940
Callao	12°04.0′S	077°09.0′W	—	12:08:04.9			21	102	202	299	0.950	0.946
Chiclayo	06°46.0′S	079°51.0′W	—	12:02:00.9			16	104	201	294	0.831	0.796
Chimbote	09°05.0′S	078°36.0′W	—	12:04:30.5			18	103	201	296	0.883	0.862
Cuzco	13°31.0′S	071°59.0′W	3565	12:12:15.5			27	100	203	300	0.929	0.921
Guerreros Estn.	16°52.0′S	072°01.3′W	374	12:16:25.3	02m49.1s	169	29	98	203	304	1.045	1.000
Pachia	18°54.6′S	070°09.1′W	394	12:20:24.9			32	96	23	126	0.996	0.998
Huancayo	12°04.0′S	075°14.0′W	—	12:08:54.8			23	101	202	299	0.928	0.919
Ica	14°04.0′S	075°42.0′W	—	12:10:57.2			24	101	202	301	0.988	0.991
Iquitos	03°46.0′S	073°15.0′W	—	12:01:39.5			22	105	202	290	0.670	0.595
Juliaca	15°30.0′S	070°08.0′W	—	12:15:57.0			30	99	203	302	0.960	0.959
Lima	12°03.0′S	077°03.0′W	129	12:08:06.1			21	102	202	299	0.948	0.945
Mollendo	17°02.0′S	072°01.0′W	—	12:16:38.7	02m51.2s	170	29	98	23	124	1.045	1.000
Moquegua	17°12.0′S	070°56.0′W	—	12:17:35.0	02m42.6s	171	30	98	203	304	1.045	1.000
Nazca	14°50.0′S	074°57.0′W	—	12:12:13.4			25	100	202	302	1.000	1.000
Piura	05°12.0′S	080°38.0′W	—	12:00:30.4			15	104	201	293	0.794	0.749
Pucallpa	08°23.0′S	074°32.0′W	—	12:05:22.1			22	103	202	295	0.818	0.779
Sullana	04°53.0′S	080°41.0′W	—	12:00:14.3			14	104	201	292	0.786	0.738
Tacna	18°01.0′S	070°15.0′W	—	12:19:08.7	02m53.1s	172	31	97	23	125	1.046	1.000
Tarata	17°28.0′S	070°02.0′W	—	12:18:34.3	02m28.5s	172	31	97	203	304	1.046	1.000
Trujillo	08°07.0′S	079°02.0′W	—	12:03:27.5			17	103	201	295	0.861	0.833
Vitarte	12°02.0′S	076°56.0′W	—	12:08:08.0			21	102	202	299	0.946	0.942
URUGUAY												
Montevideo	34°53.0′S	056°11.0′W	24	13:00:11.4			51	69	23	151	0.782	0.735
Paysandú	32°19.0′S	058°05.0′W	—	12:53:30.9			49	75	23	146	0.820	0.783
Rivera	30°54.0′S	055°31.0′W	—	12:54:43.0			52	74	24	145	0.888	0.870
Salto	31°23.0′S	057°58.0′W	—	12:52:09.4			49	76	24	144	0.844	0.814
FRENCH GUIANA												
Cayenne	04°56.0′N	052°20.0′W	7	12:11:01.5			41	115	205	275	0.125	0.053
GUYANA												
Georgetown	06°48.0′N	058°10.0′W	2	12:03:34.1			33	113	204	276	0.157	0.074
SURINAME												
Paramaribo	05°50.0′N	055°10.0′W	4	12:07:08.6			37	114	205	275	0.142	0.064

Table 11b
LOCAL CIRCUMSTANCES DURING THE TOTAL SOLAR ECLIPSE OF 3 NOVEMBER 1994 FOR PARAGUAY, PERU AND URUGUAY

Location Name	First Contact U.T. h m s	Alt °	P °	V °	Second Contact U.T. h m s	Alt °	P °	V °	Third Contact U.T. h m s	Alt °	P °	V °	Fourth Contact U.T. h m s	Alt °	P °	V °
PARAGUAY																
Asuncion	11:32:32.4	33	294	44	12:42:20.3	49	192	303	12:43:01.0	49	214	325	14:01:43.9	66	112	230
Caaguazu	11:33:35.7	35	292	42	12:43:09.8	51	93	204	12:46:39.7	51	313	64	14:05:15.8	69	114	234
Concepcion	11:30:05.3	33	291	39									13:59:46.3	67	115	229
Encarnación	11:37:07.6	35	295	48									14:08:11.8	68	110	237
Fernando d.l....	11:32:39.9	33	294	44	12:42:40.8	49	198	309	12:43:00.8	49	208	320	14:01:56.3	66	112	231
Marsiscal	11:25:53.2	28	291	38	12:32:49.0	44	66	172	12:35:11.3	44	340	86	13:51:11.9	62	115	223
Rosario	11:31:45.7	33	292	42	12:40:27.2	49	102	212	12:44:02.6	50	304	54	14:01:47.8	67	114	231
Villarrica	11:34:11.9	34	294	45	12:43:59.8	50	166	278	12:46:14.2	51	240	353	14:04:55.8	68	112	233
PERU																
Arequipa	11:14:39.3	15	290	33	12:15:38.2	29	42	142	12:16:35.8	29	2	103	13:25:33.0	46	114	212
Ayacucho	11:11:02.6	11	286	27									13:17:48.1	41	117	211
Callao	11:09:57.8	7	286	27									13:13:24.5	37	116	210
Chiclayo	11:06:29.7	3	279	16									13:04:08.5	31	123	211
Chimbote	11:07:45.5	5	282	20									13:08:10.2	34	120	211
Cuzco	11:11:40.6	13	286	27									13:20:50.7	44	118	212
Guerreros Est...	11:15:07.5	15	291	35	12:15:01.2	29	103	204	12:17:50.3	29	301	42	13:25:37.5	45	113	212
Pachia	11:17:54.9	17	293	38									13:30:59.0	49	111	213
Huancayo	11:10:02.8	9	285	25									13:15:16.3	39	118	211
Ica	11:11:48.7	10	289	31									13:17:31.0	40	114	210
Iquitos	11:06:10.3	8	271	3									13:04:11.3	37	133	216
Juliaca	11:14:02.5	16	288	30									13:26:07.5	47	117	213
Lima	11:09:56.9	8	286	27									13:13:28.7	37	116	210
Mollendo	11:15:19.2	15	291	35	12:15:13.4	29	114	215	12:18:04.6	29	290	31	13:25:52.0	45	113	212
Moquegua	11:15:42.5	16	291	35	12:16:14.4	30	91	192	12:18:57.0	31	313	54	13:27:31.4	47	113	212
Nazca	11:12:36.9	11	289	32									13:19:21.2	41	114	210
Piura	11:05:51.8	2	277	12									13:01:31.6	29	125	212
Pucallpa	11:07:34.7	8	279	16									13:10:34.9	38	124	213
Sullana	11:05:45.1	1	277	11									13:01:04.0	29	125	212
Tacna	11:16:48.4	17	292	36	12:17:42.2	31	124	226	12:20:35.4	32	280	22	13:29:36.3	48	113	213
Tarata	11:16:12.6	17	291	35	12:17:20.9	31	79	181	12:19:49.4	32	325	67	13:29:07.5	48	114	213
Trujillo	11:07:10.9	4	281	18									13:06:32.3	33	121	211
Vitarte	11:09:56.3	8	286	27									13:13:34.1	37	117	210
URUGUAY																
Montevideo	11:50:44.3	38	305	68									14:15:45.1	64	98	242
Paysandú	11:44:28.8	35	303	62									14:09:25.2	64	101	235
Rivera	11:43:34.6	37	300	57									14:13:18.2	67	105	241
Salto	11:42:47.1	35	302	60									14:08:41.8	64	102	235
FRENCH GUIANA																
Cayenne	11:37:18.5	33	233	307									12:49:18.9	49	176	240
GUYANA																
Georgetown	11:28:07.3	25	235	310									12:43:12.6	42	172	238
SURINAME																
Paramaribo	11:32:22.8	29	234	309									12:46:18.6	46	174	239

Table 12a
CIRCUMSTANCES AT MAXIMUM ECLIPSE ON 3 NOVEMBER 1994
FOR MEXICO, CENTRAL AMERICA AND THE CARIBBEAN

Location Name	Latitude	Longitude	Elev. m	U.T. of Maximum Eclipse h m s	Umbral Duration	Path Width km	Sun Alt °	Sun Azm °	P °	V °	Eclipse Mag.	Eclipse Obs.
MEXICO												
Campeche	19°51.0'N	090°32.0'W	–	12:05 Rise			0	106	–	–	0.079	0.026
Coatzacoalcos	18°09.0'N	094°25.0'W	–	12:19 Rise			0	106	–	–	0.056	0.016
Merida	20°58.0'N	089°37.0'W	24	12:02 Rise			0	106	–	–	0.051	0.014
Minatitlan	17°59.0'N	094°31.0'W	–	12:19 Rise			0	106	–	–	0.060	0.018
Salina Cruz	16°11.0'N	095°12.0'W	60	12:18 Rise			0	105	–	–	0.119	0.049
Tuxtla Gutierrez	16°45.0'N	093°07.0'W	–	12:12 Rise			0	106	–	–	0.147	0.067
BELIZE												
Belize City	17°30.0'N	088°12.0'W	6	11:53 Rise			0	106	–	–	0.170	0.083
Belmopan	17°15.0'N	088°46.0'W	–	11:55 Rise			0	106	–	–	0.180	0.090
COSTA RICA												
Cartago	09°51.0'N	083°55.0'W	–	11:51:57.1			5	106	200	278	0.373	0.261
Limon	09°59.0'N	083°01.0'W	–	11:51:54.6			6	107	200	278	0.360	0.248
Puntarenas	09°58.0'N	084°50.0'W	–	11:51:56.8			4	106	200	278	0.378	0.266
San Jose	09°56.0'N	084°05.0'W	1234	11:51:54.7			5	106	200	278	0.372	0.260
EL SALVADOR												
San Miguel	13°28.0'N	088°10.0'W	–	11:51:33.4			1	106	199	275	0.298	0.189
San Salvador	13°42.0'N	089°12.0'W	734	11:51:40.5			0	105	199	275	0.300	0.190
Santa Ana	14°00.0'N	089°33.0'W	–	11:54 Rise			0	105	–	–	0.291	0.182
GUATEMALA												
Antigua	14°33.0'N	090°42.0'W	–	12:00 Rise			0	105	–	–	0.269	0.162
Mazatenango	14°31.0'N	091°30.0'W	–	12:03 Rise			0	105	–	–	0.262	0.156
Quezaltenango	14°51.0'N	091°31.0'W	–	12:03 Rise			0	105	–	–	0.250	0.145
HONDURAS												
San Pedro Sula	15°27.0'N	088°02.0'W	–	11:51:20.5			0	106	199	273	0.235	0.133
Tegucigalpa	14°06.0'N	087°13.0'W	–	11:51:21.2			1	106	199	275	0.270	0.163
NICARAGUA												
Bluefields	12°00.0'N	083°49.0'W	–	11:51:26.0			5	106	200	276	0.305	0.195
Granada	11°56.0'N	085°58.0'W	–	11:51:33.7			3	106	200	277	0.327	0.215
Leon	12°25.0'N	086°53.0'W	–	11:51:33.5			2	106	199	276	0.320	0.209
Managua	12°09.0'N	086°17.0'W	–	11:51:32.8			3	106	200	276	0.323	0.212
PANAMA												
Colon	09°22.0'N	079°54.0'W	–	11:52:15.2			9	107	201	278	0.349	0.236
David	08°26.0'N	082°26.0'W	–	11:52:23.9			7	106	200	280	0.403	0.291
Panama City	08°58.0'N	079°31.0'W	–	11:52:25.5			10	107	201	279	0.357	0.244
San Miguelito	11°24.0'N	084°54.0'W	–	11:51:36.1			4	106	200	277	0.334	0.222
ARUBA												
Oranjestad	12°33.0'N	070°06.0'W	–	11:53:34.1			18	110	202	273	0.141	0.063
GRENADA												
St. George's	12°03.0'N	061°45.0'W	–	11:58:06.2			26	114	204	272	0.050	0.013
JAMAICA												
Kingston	18°00.0'N	076°48.0'W	36	11:51:06.9			9	109	201	269	0.048	0.012
NETHERLANDS ANTILLES												
Willemstad, Curacao	12°06.0'N	068°56.0'W	–	11:54:09.5			19	111	202	274	0.141	0.063
Kralendijk, Bonaire	12°09.0'N	068°16.0'W	–	11:54:25.7			20	111	202	273	0.131	0.056
TRINIDAD & TOBAGO												
Port-of-Spain	10°39.0'N	061°31.0'W	22	11:58:49.7			27	113	204	273	0.089	0.032
San Fernando	10°17.0'N	061°28.0'W	–	11:59:01.6			28	113	204	273	0.099	0.037

Table 12b
LOCAL CIRCUMSTANCES DURING THE TOTAL SOLAR ECLIPSE OF 3 NOVEMBER 1994 FOR MEXICO, CENTRAL AMERICA AND THE CARIBBEAN

Location Name	First Contact U.T. h m s	Alt °	P °	V °	Second Contact U.T. h m s	Alt °	P °	V °	Third Contact U.T. h m s	Alt °	P °	V °	Fourth Contact U.T. h m s	Alt °	P °	V °
MEXICO																
Campeche	–												12:17:04.1	2	172	241
Coatzacoalcos	–												12:24:33.5	1	162	233
Merida	–												12:12:01.6	2	177	245
Minatitlan	–												12:25:00.1	1	162	233
Salina Cruz	–												12:29:08.2	2	156	229
Tuxtla Gutier...	–												12:26:54.0	3	159	231
BELIZE																
Belize City	–												12:22:12.6	6	166	236
Belmopan	–												12:23:13.8	6	165	235
COSTA RICA																
Cartago	–												12:37:20.3	16	150	225
Limon	–												12:36:58.2	17	151	226
Puntarenas	–												12:37:14.7	15	150	225
San Jose	–												12:37:11.1	16	150	225
EL SALVADOR																
San Miguel	–												12:31:25.1	9	155	228
San Salvador	–												12:31:18.8	8	155	228
Santa Ana	–												12:30:52.1	8	155	228
GUATEMALA																
Antigua	–												12:30:14.6	7	156	229
Mazatenango	–												12:30:38.2	6	155	228
Quezaltenango	–												12:29:59.9	6	156	229
HONDURAS																
San Pedro Sula	–												12:27:10.0	8	160	232
Tegucigalpa	–												12:29:46.3	10	157	230
NICARAGUA																
Bluefields	–												12:33:05.1	14	155	228
Granada	–												12:33:45.8	13	153	227
Leon	–												12:33:04.3	11	153	227
Managua	–												12:33:25.5	12	153	227
PANAMA																
Colon	11:10:18.6	0	249	329									12:37:52.4	20	152	226
David	–												12:39:50.5	18	148	224
Panama City	11:09:58.5	0	249	330									12:38:40.3	21	152	226
San Miguelito	–												12:34:33.0	14	153	227
ARUBA																
Oranjestad	11:23:05.0	11	232	306									12:26:30.5	25	172	240
GRENADA																
St. George's	11:38:18.4	22	221	291									12:19:51.5	31	185	251
JAMAICA																
Kingston	11:33:43.3	5	218	288									12:09:14.9	13	183	250
NETHERLANDS ANTILLES																
Willemstad, C...	11:23:25.8	12	232	306									12:27:27.5	27	172	240
Kralendijk, B...	11:24:32.6	13	231	305									12:26:49.4	27	173	241
TRINIDAD & TOBAGO																
Port-of-Spain	11:32:32.3	21	227	299									12:27:44.6	34	179	245
San Fernando	11:31:18.0	21	228	301									12:29:33.0	34	178	244

Table 13a
CIRCUMSTANCES AT MAXIMUM ECLIPSE ON 3 NOVEMBER 1994
FOR AFRICA

Location Name	Latitude	Longitude	Elev. m	U.T. of Maximum Eclipse h m s	Umbral Duration	Path Width km	Sun Alt °	Sun Azm °	P °	V °	Eclipse Mag.	Eclipse Obs.
BOTSWANA												
Francistown	21°11.0′S	027°32.0′E	1081	15:21:26.4			13	258	187	78	0.549	0.451
Gaborone	24°45.0′S	025°55.0′E	—	15:18:49.4			15	260	187	75	0.649	0.569
KENYA												
Mombasa	04°03.0′S	039°40.0′E	17	15:11 Set			0	255	—	—	0.065	0.020
LESOTHO												
Maseru	29°28.0′S	027°30.0′E	—	15:16:16.4			15	261	188	71	0.805	0.763
MADAGASCAR												
Antananarivo	18°55.0′S	047°31.0′E	—	14:56 Set			0	254	—	—	0.394	0.281
MALAWI												
Lilongwe	13°59.0′S	033°44.0′E	—	15:25:42.5			4	255	186	83	0.372	0.259
MOZAMBIQUE												
Beira	19°49.0′S	034°52.0′E	9	15:23:39.4			5	256	186	77	0.564	0.468
Maputo	25°58.0′S	032°35.0′E	64	15:20:00.2			10	258	187	72	0.735	0.674
NAMIBIA												
Windhoek	22°34.0′S	017°06.0′E	1860	15:14:38.9			24	263	188	80	0.522	0.421
SOUTH AFRICA												
Bloemfontein	29°12.0′S	026°07.0′E	—	15:15:56.6			17	262	188	71	0.788	0.741
Cape Town	33°55.0′S	018°22.0′E	18	15:08:01.5			25	268	189	68	0.886	0.866
Durban	29°48.0′S	031°00.0′E	5	15:17:02.7			12	259	187	69	0.839	0.806
Johannesburg	26°08.0′S	027°54.0′E	—	15:18:42.6			14	260	187	74	0.706	0.639
Kimberley	28°40.0′S	024°50.0′E	1288	15:15:47.8			18	262	188	72	0.763	0.710
Port Elizabeth	33°58.0′S	025°40.0′E	62	15:12:00.4			18	264	188	67	0.929	0.920
Pretoria	25°45.0′S	028°10.0′E	1473	15:19:04.1			14	260	187	74	0.696	0.627
Vereeniging	26°38.0′S	027°57.0′E	—	15:18:24.1			14	260	187	73	0.722	0.658
SWAZILAND												
Mbabane	28°18.0′S	031°06.0′E	—	15:18:09.0			12	259	187	71	0.795	0.749
TANZANIA												
Dar-es-Salaam	06°48.0′S	039°17.0′E	15	15:16 Set			0	255	—	—	0.168	0.081
Dodoma	06°11.0′S	035°45.0′E	—	15:27:07.6			1	255	186	89	0.142	0.064
Arusha	03°22.0′S	036°41.0′E	—	15:22 Set			0	255	—	—	0.057	0.016
Zanzibar	06°10.0′S	039°11.0′E	—	15:15 Set			0	255	—	—	0.144	0.065
ZAIRE												
Kananga	06°14.0′S	022°17.0′E	—	15:22:42.4			14	256	187	94	0.032	0.007
Lubumbashi	11°40.0′S	027°28.0′E	—	15:24:46.4			10	256	187	87	0.247	0.143
Mbuji-Mayi	06°09.0′S	023°38.0′E	—	15:23:32.1			13	256	187	94	0.040	0.010
ZAMBIA												
Kitwe	12°49.0′S	028°13.0′E	—	15:24:47.0			10	257	186	86	0.290	0.181
Livingstone	17°50.0′S	025°53.0′E	—	15:22:17.0			13	258	187	82	0.431	0.321
Lusaka	15°25.0′S	028°17.0′E	1375	15:24:05.5			10	257	187	83	0.373	0.260
Ndola	12°58.0′S	028°38.0′E	—	15:24:53.0			9	256	186	85	0.298	0.188
ZIMBABWE												
Bulawayo	20°09.0′S	028°36.0′E	1445	15:22:18.3			11	258	187	79	0.525	0.424
Harare	17°50.0′S	031°03.0′E	1585	15:23:58.2			8	257	186	80	0.471	0.364
ST. HELENA												
Gough Is.	40°20.0′S	010°00.0′W	—	14:29:09.7	03m46.2s	180	53	304	15	237	1.051	1.000

Table 13b
LOCAL CIRCUMSTANCES DURING THE TOTAL SOLAR ECLIPSE OF 3 NOVEMBER 1994
FOR AFRICA

Location Name	First Contact U.T. h m s	Alt °	P °	V °	Second Contact U.T. h m s	Alt °	P °	V °	Third Contact U.T. h m s	Alt °	P °	V °	Fourth Contact U.T. h m s	Alt °	P °	V °
BOTSWANA																
Francistown	14:25:30.0	26	250	143									16:12:17.4	2	124	12
Gaborone	14:19:06.2	29	256	146									16:12:50.8	4	118	3
KENYA																
Mombasa	15:02:03.9	2	212	119									-			
LESOTHO																
Maseru	14:14:21.9	29	266	150									16:12:15.4	4	109	349
MADAGASCAR																
Antananarivo	14:32:31.7	5	254	146									-			
MALAWI																
Lilongwe	14:39:56.2	15	237	136									-			
MOZAMBIQUE																
Beira	14:30:41.1	17	250	144									-			
Maputo	14:21:45.1	22	261	149									-			
NAMIBIA																
Windhoek	14:13:59.8	38	249	142									16:09:31.4	12	126	16
SOUTH AFRICA																
Bloemfontein	14:13:33.9	30	265	150									16:12:17.4	5	110	351
Cape Town	14:00:27.7	39	272	150									16:08:51.4	13	105	343
Durban	14:16:27.2	25	267	151									16:12:00.5	1	107	346
Johannesburg	14:18:42.6	27	260	148									16:13:00.7	2	114	358
Kimberley	14:13:07.4	31	263	149									16:12:21.2	6	111	353
Port Elizabeth	14:08:02.4	32	273	152									16:09:52.5	7	102	339
Pretoria	14:19:24.8	27	259	148									16:13:04.6	2	115	359
Vereeniging	14:18:06.5	27	261	148									16:12:57.6	2	114	357
SWAZILAND																
Mbabane	14:18:11.2	25	265	150									16:12:32.3	1	109	350
TANZANIA																
Dar-es-Salaam	14:54:14.7	5	222	126									-			
Dodoma	14:57:55.3	7	216	122									-			
Arusha	15:07:55.3	3	206	113									-			
Zanzibar	14:55:57.9	4	219	124									-			
ZAIRE																
Kananga	15:07:11.1	18	201	109									15:38:40.4	10	172	78
Lubumbashi	14:44:14.8	20	227	130									16:02:41.1	1	145	43
Mbuji-Mayi	15:06:22.5	17	203	111									15:40:55.6	8	170	76
ZAMBIA																
Kitwe	14:41:34.4	20	231	133									16:04:57.4	1	142	38
Livingstone	14:30:14.1	26	242	139									16:09:54.0	3	131	24
Lusaka	14:36:01.6	22	237	137									16:08:21.3	0	135	29
Ndola	14:41:18.8	19	232	133									16:05:21.1	0	141	37
ZIMBABWE																
Bulawayo	14:27:44.4	24	248	142									16:12:00.4	0	125	14
Harare	14:32:35.4	20	244	140									-			
ST. HELENA																
Gough Is.	13:08:57.1	63	286	125	14:27:15.7	53	117	338	14:31:01.9	53	270	132	15:43:55.9	40	102	332

Table 14
CLIMATOLOGICAL STATISTICS ALONG THE ECLIPSE TRACK

Station	Days with clear skies and good visibilities (1200 UTC)	Mean Cloud Cover (tenths)	Sunshine hours in November	Days with low cloud and/or fog	Monthly Rainfall (inches)	Prevailing Wind
Pisco, Peru	7.0	-	-	5.9	0.10	-
San Juan, Peru	9.8	-	-	6.3	0.19	NE
Arequipa, Peru	20.8	-	-	0.0	0.03	W
Tacna, Peru	3.8	-	-	15.7	0.04	S
Cuzco, Peru	7.1	-	-	0.7	3.00	-
Arica, Chile	4.4	4.5	-	2.3	0.00	SW
Iquique, Chile	7.2	-	-	-	0.00	-
Charaña, Bolivia	18.9	-	-	0.2	0.35	-
Uyuni, Bolivia	23.7	-	-	0.0	0.20	-
Sucre, Bolivia	6.3	-	-	5.6	2.60	-
La Quiaca, Argentina	13.4	4.3	300	0.0	1.00	NE
Tarija, Bolivia	8.1	-	-	2.4	1.90	-
Camiri, Bolivia	9.8	-	-	3.7	3.32	-
Yacuiba, Bolivia	7.4	-	-	5.6	3.57	S
Rivadavia, Argentina	10.5	-	-	18.0	2.35	-
Mariscal, Paraguay	8.6	5	-	2.6	3.31	N/NE
Gran Chaco, Paraguay	-	3.3	264	-	-	NE
Puerto Casado, Paraguay	-	5	-	-	-	S
Concepción, Paraguay	16.7	-	-	3.3	-	-
Asunción, Paraguay	10.1	4	270	1.9	5.90	E
Formosa, Argentina	16.7	4.4	264	0.0	6.00	E
Villarrica, Paraguay	24.0	-	-	6.0	5.21	-
Pilar, Paraguay	10.0	-	-	10.0	5.51	-
San Juan Bautista, Paraguay	13.2	-	-	1.8	6.75	-
Posadas, Argentina	10.6	3.8	237	6.7	5.03	S
Iguassu Falls, Argentina	-	5	-	-	-	SE
Alegrete, Brazil	-	4.2	258	-	-	E/SE
Curitiba, Brazil	2.5	7.1	178	19.4	5.00	E
Porto Alegre, Brazil	9.3	5.1	234	0.0	3.10	SE
Florianapolis, Brazil	5.4	-	-	2.4	3.50	-
Gough Island, UK	-	8.0	122	-	-	W

Table 15

SOLAR ECLIPSE EXPOSURE GUIDE

ISO					f/Number				
25	1.4	2	2.8	4	5.6	8	11	16	22
50	2	2.8	4	5.6	8	11	16	22	32
100	2.8	4	5.6	8	11	16	22	32	44
200	4	5.6	8	11	16	22	32	44	64
400	5.6	8	11	16	22	32	44	64	88
800	8	11	16	22	32	44	64	88	128
1600	11	16	22	32	44	64	88	128	176

Subject	Q				Shutter Speed				
Solar Eclipse									
Partial - 4.0 ND	11	—	—	—	1/4000	1/2000	1/1000	1/500	1/250	1/125
Partial - 5.0 ND	8	1/4000	1/2000	1/1000	1/500	1/250	1/125	1/60	1/30	1/15
Bailey's Beads[1]	12	—	—	—	—	1/4000	1/2000	1/1000	1/500	1/250
Chromosphere	11	—	—	—	1/4000	1/2000	1/1000	1/500	1/250	1/125
Prominences	9	—	1/4000	1/2000	1/1000	1/500	1/250	1/125	1/60	1/30
Corona - 0.1 Rs	7	1/2000	1/1000	1/500	1/250	1/125	1/60	1/30	1/15	1/8
Corona - 0.2 Rs[2]	5	1/500	1/250	1/125	1/60	1/30	1/15	1/8	1/4	1/2
Corona - 0.5 Rs	3	1/125	1/60	1/30	1/15	1/8	1/4	1/2	1 sec	2 sec
Corona - 1.0 Rs	1	1/30	1/15	1/8	1/4	1/2	1 sec	2 sec	4 sec	8 sec
Corona - 2.0 Rs	0	1/15	1/8	1/4	1/2	1 sec	2 sec	4 sec	8 sec	15 sec
Corona - 4.0 Rs	-1	1/8	1/4	1/2	1 sec	2 sec	4 sec	8 sec	15 sec	30 sec
Corona - 8.0 Rs	-3	1/2	1 sec	2 sec	4 sec	8 sec	15 sec	30 sec	1 min	2 min

Exposure Formula: $t = f^2 / (I \times 2^Q)$ where: t = exposure time (sec)
f = focal ratio
I = ISO film speed
Q = brightness exponent

Abbreviations: ND = Neutral Density Filter.
Rs = Solar Radii.

Notes: [1] Bailey's Beads are extremely bright and change rapidly.
[2] This exposure also recommended for the 'Diamond Ring' effect.

F. Espenak - Nov 1992

TOTAL SOLAR ECLIPSE OF 3 NOVEMBER 1994

MAPS OF THE UMBRAL PATH

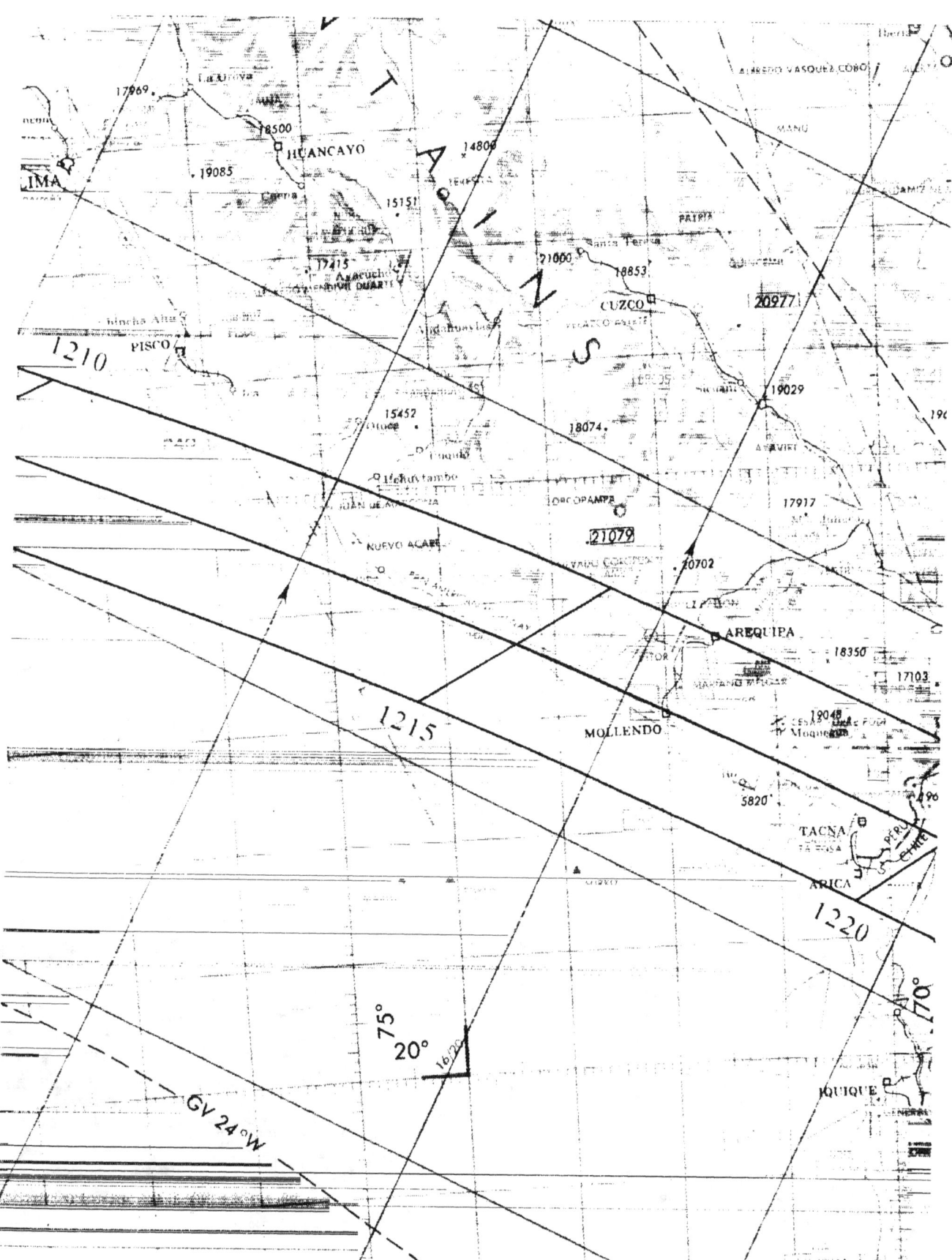

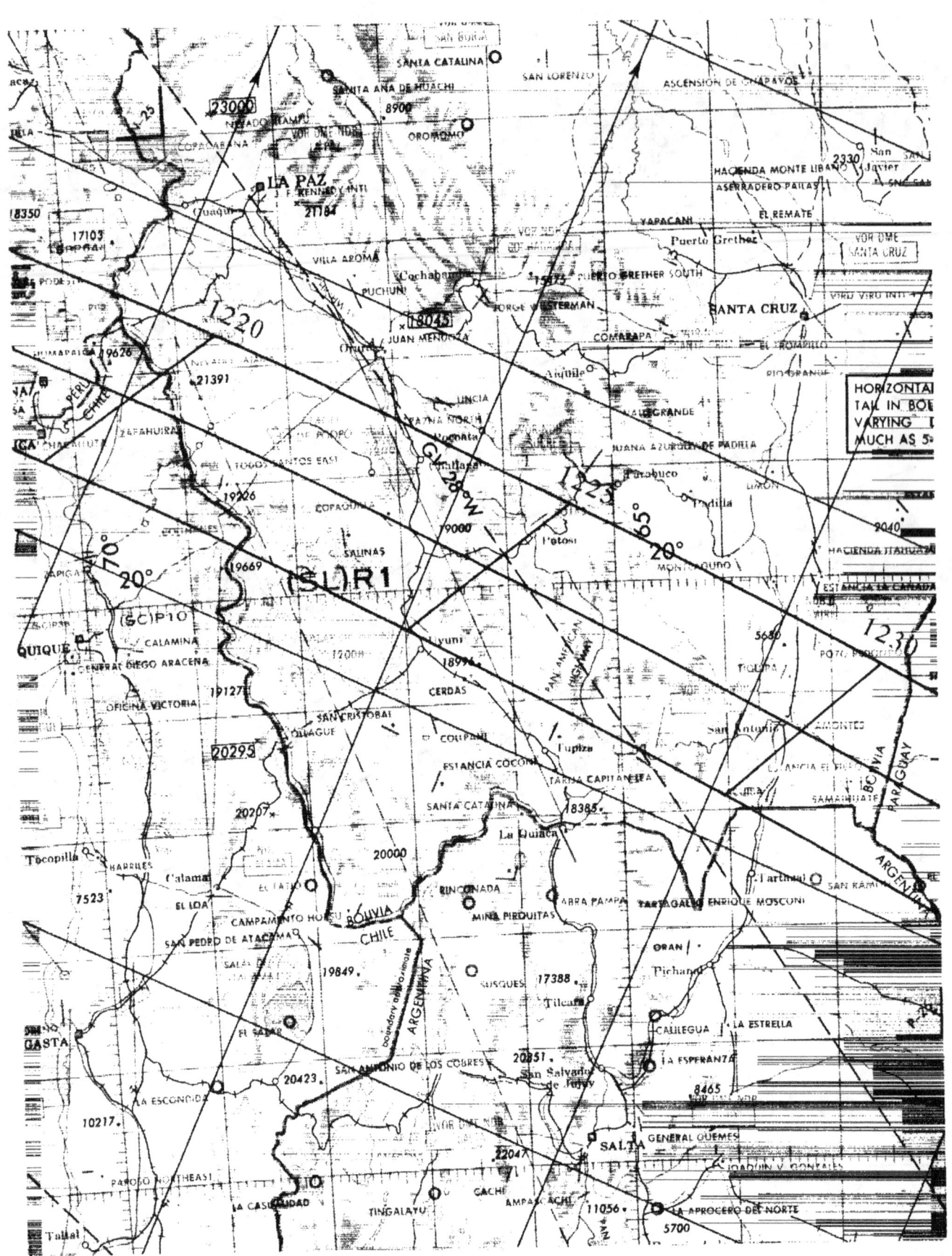

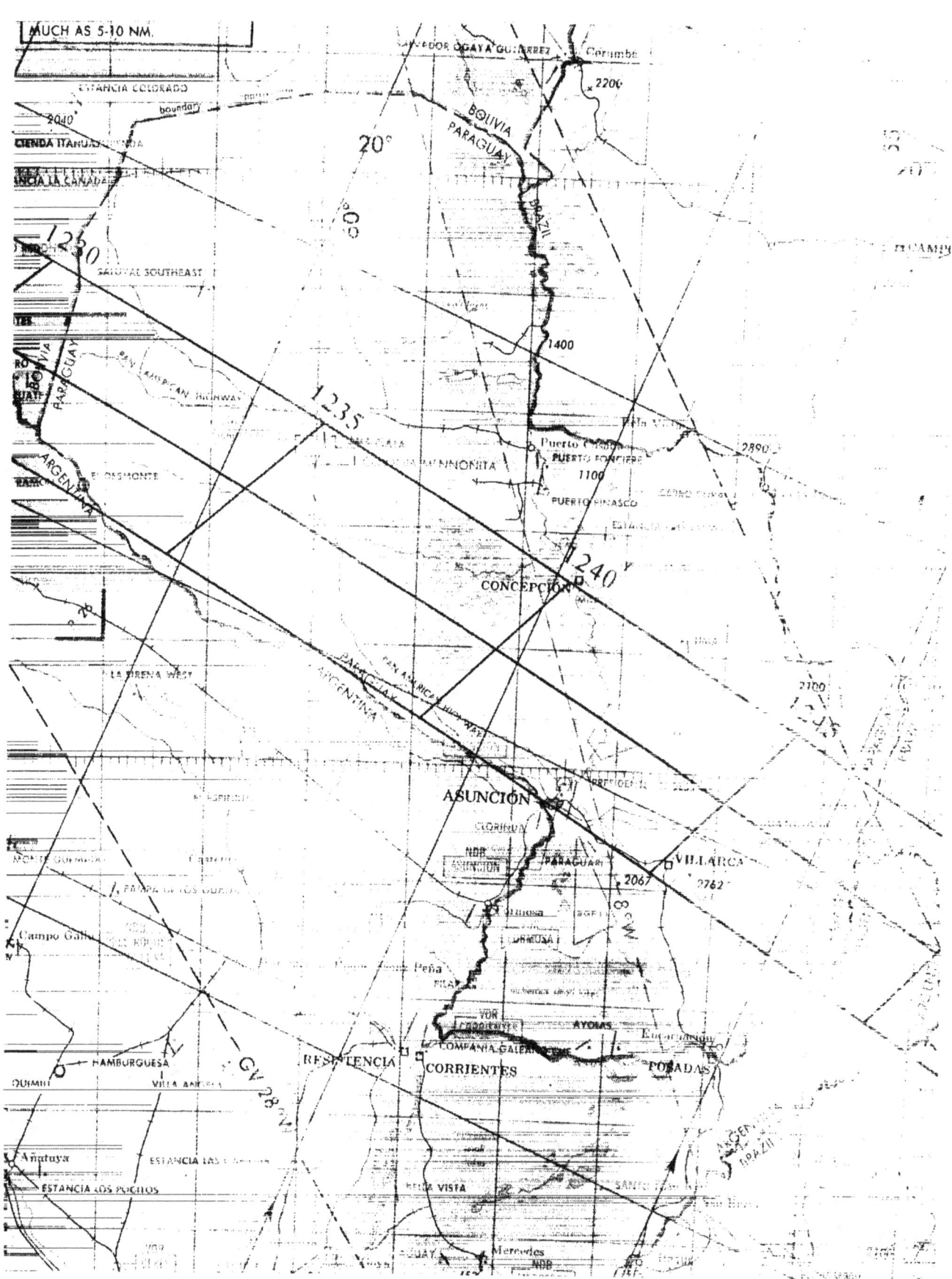

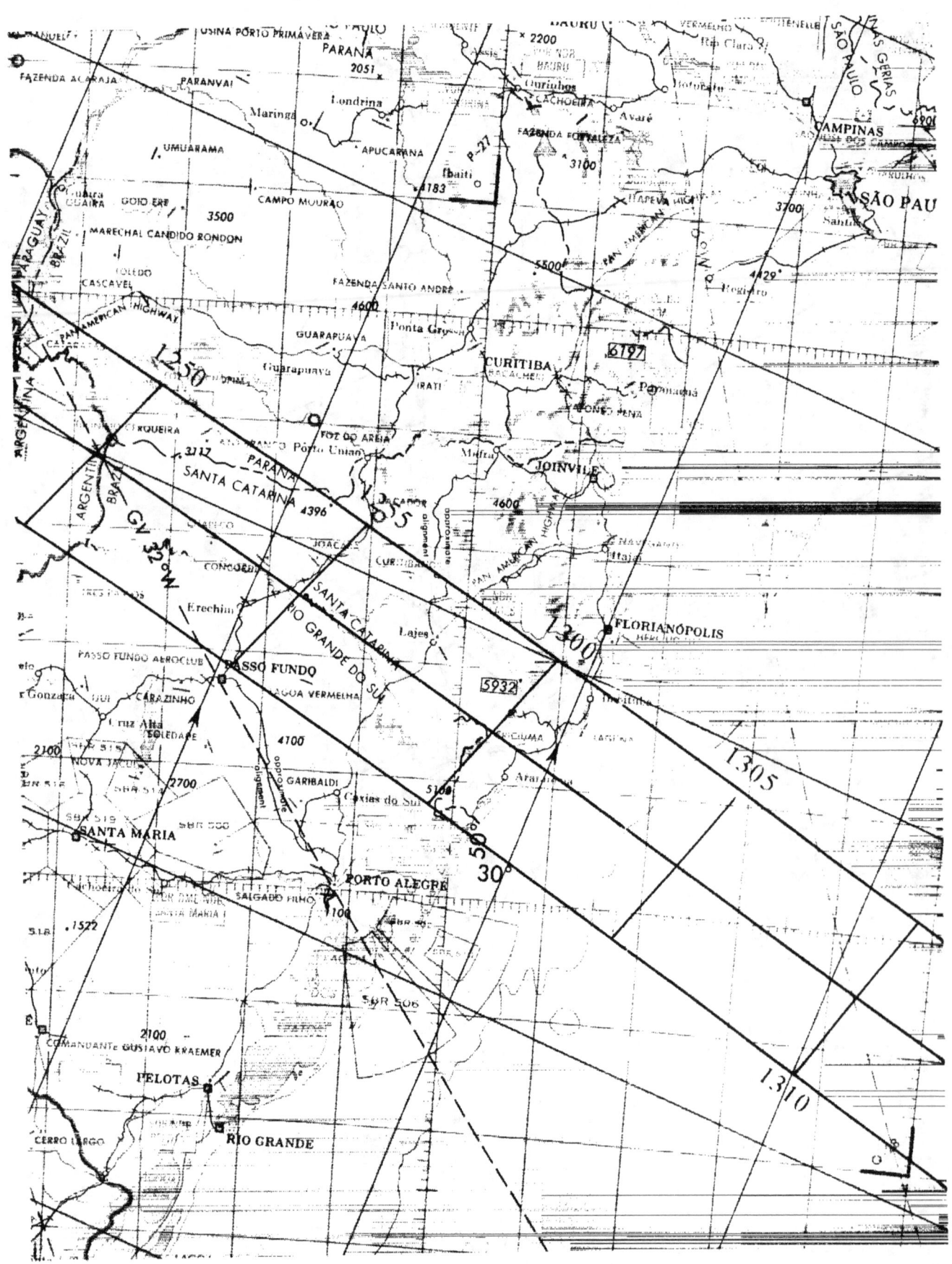

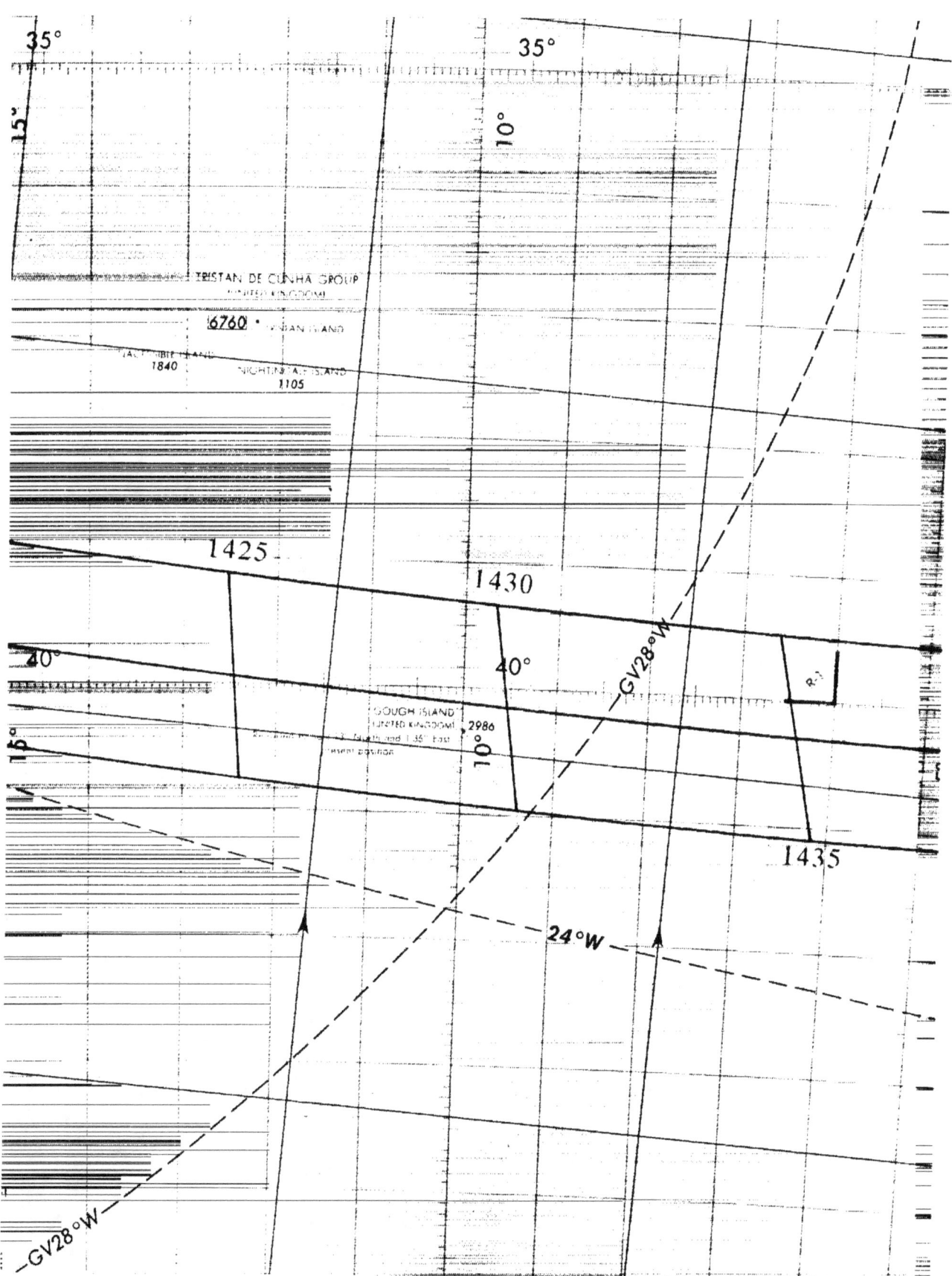

www.ingramcontent.com/pod-product-compliance
Lightning Source LLC
Chambersburg PA
CBHW081737170526
45167CB00009B/3846